ENVIRONMENTAL REMEDIATION TECHNOLOGIES, REGULATIONS AND SAFETY

TROPICAL FOREST, GEOSPATIAL DATA AND REDD+

ENVIRONMENTAL REMEDIATION TECHNOLOGIES, REGULATIONS AND SAFETY

Additional books in this series can be found on Nova's website under the Series tab.

Additional e-books in this series can be found on Nova's website under the eBooks tab.

ENVIRONMENTAL REMEDIATION TECHNOLOGIES, REGULATIONS AND SAFETY

TROPICAL FOREST, GEOSPATIAL DATA AND REDD+

RAM AVTAR

nova science publishers
New York

Library of Congress Cataloging-in-Publication Data

ISBN: 978-1-63485-277-7

Published by Nova Science Publishers, Inc. † New York

CONTENTS

LIST OF TABLES

LIST OF FIGURES

PREFACE

This is an attempt to identify the basic principles that should most cost effective and potential means of reducing the emission by implementation of REDD+ (Reducing emissions from deforestation and forest degradation). The views it sets out are rooted in the conviction that was signified by 15th meet of COP in Copenhagen in 2009, where REDD+ regarded as an effective strategy to alleviate poverty, reducing deforestation and degradation of forest, sustainable forest management and enhancement of carbon stocks to mitigate climate change and also socioeconomic development of developing countries. This book also emphasis on an urgent need of establishing measurement, reporting and verification (MRV) system for accurate estimation of forest carbon stocks.

The purposes of the present strategy of REDD+ are three-fold and carried out in three phases: To lay out the readiness phase; demonstration phase and result action based. Most of the countries worldwide is in initial phase only and UNFCCC is helping them to get ready for the full-fledged implementation of REDD+. The book explained the how these countries which are participating in the REDD+ mitigation mechanism under UNFCCC, have to establish two key points as reference emission level and MRV system for monitoring forests.

The necessity of such book is driven by a sense of urgency among all the scientific community that we have a limited window of understanding and opportunity to REDD+ and exploring its potential. Further, understanding of the importance and role of REDD+ mechanisms to manage post-Kyoto climate change regime. With the chapters in the book we understand the related key concepts about over all forest distribution, management and the core concept of REDD+ mechanism. Next chapter prominently took the study area which is

Cambodia in this regards and forest management plan for the government and the related activities to be implemented in order to establish the national REDD+ program in the country. Also focusing on the national readiness roadmap and how this will be helpful to reduce greenhouse gas emissions, alleviating poverty in the country, improves forest governance and enhancing sustainable management in Cambodia.

What our book does insist upon is the critical need for our study area, the country where all related studies are done and in efforts of making to widely useful the next chapter detailed about use of full polarimetric PALSAR data and polarmetric decomposition model to monitor different types of forest cover. Also experiments and classification showed improvement of better forest mapping and management practices due to use of polarmetric data in detection of structural differences between forest canopies. Next chapter introduce the characterization of forest and deforestation using PALSAR data based on backscattering and polarmetric parameters of data, is to identify the deforested area during rainy season when optical sensor cannot provide such images. Chapter 5 presents the use of various DEMs data for estimating the forest height of cut over forest area. The accurate estimation of forest volumes is important to understand and underlies in this chapter where forest height is an important parameter. The following chapter detailed quantitative analysis of data and explains about the advantage and disadvantages of traditional techniques of collecting data and optical sensors limitation are documented well enough to understand the importance of PALSAR mosaic data for national level forest biomass estimation and emphasis given on research on integration of multi source data and remote sensing, GIS and modeling techniques. The next chapter closely scrutinizes the role of remote sensing techniques and community forestry for effective forest management in Cambodia. The given analysis in the chapter explains the real need and how these techniques can be important in REDD+. The last chapter of the book analyses the quantitative work and draw conclusion about each one chapter and also suggest the author recommendation and future work scope which gives a good inspiration for younger researchers to look and focus upon.

This book "Tropical Forest, Geospatial Data and REDD+" is the product of collective deliberation and long research work by author (Ram Avtar) and guidance and help of Prof. Haruo Sawada. Also this book wouldn't be book without exceptional help and commitment of people from Cambodia. In summary this book is an introduction of MRV systems and its establishment to strengthen REDD+ mechanism. This book will provide the motivations for the further research work in the field of climate change mitigation and the limitations of geospatial data can be considered as future research areas. The contribution of this book will be capacity building of local people and government in order of successful REDD+ project in the developing countries.

ACKNOWLEDGMENTS

The author is highly thankful to the Monbukagakusho (MEXT) Japanese Government Fellowship to pursue research at The University of Tokyo, Japan. He would like to thank JAXA for providing ALOS data under RA. He would also like to thank the Forestry Administration (FA), Cambodia, Forestry and Forest Products Research Institute (FFPRI), Japan and Asia Pacific Network for Global Change Research (APN-GCR) for their cooperation and support to conduct this research.

Chapter 1

INTRODUCTION

1.1. BACKGROUND

Forests, which covers about 31% of the total global land area (FAO, 2011), play a significant role in the global carbon cycle. They act as a potential carbon sink and source of atmospheric carbon dioxide (CO_2) (Muukkonen, 2007). When the forests are healthy and growing the carbon is sequestrated but when the forests are destroyed, over-harvested or burned, they become a source of CO_2 resulting the climate change. Hence, afforestation, reforestation, and avoiding the deforestation are the mechanisms of tackling climate change (Hunt, 2009). Recent reports indicate that deforestation caused a loss of about 13 million hectares (ha) of tropical forests area per year between the years 2000 to 2010 (FRA, 2010 and Achard et al., 2010).

Tropical forests contain about 40% of the carbon stock in the terrestrial biomass (Dixon et al. 1994; Phillips et al. 1998; Watson et al. 2000). The most alarming global threat in the forest sector is deforestation and forest degradation. It contributes about 17% of Green House Gases (GHG) (IPCC, 2007; Gibbs et al., 2007; Schrope, 2009; Werf et al., 2009), making deforestation as a second largest source of CO_2 emissions. The rapid economic growth, increasing demand for agricultural land, forestry products, illegal logging, urbanization are the main factors for the deforestation and forest degradation (Rudel et al., 2009). Curbing the deforestation through improved forest management practices, offers one of the most cost-effective and potential means of the emission reduction (Putz et al., 2008), and can be achieved by implementation of the Reducing Emissions from Deforestation and forest Degradation (REDD+) mechanism. REDD+ mechanism not only helps to reduce emission of CO_2 from the deforestation but also protect biodiversity, forest derived ecosystem services, and improve the socio-economic condition of the local and indigenous people, who depend on forests for their livelihood (Dudley, 2010). REDD+ assists developing countries, to develop strategies for reduction of deforestation and forest degradation, alleviate poverty, and socio-economic development of indigenous people. Recently, 15th meeting of the Conference of the Parties (COP) in Copenhagen in 2009 has pointed out significance of REDD+ as well as efforts for forest conservation, sustainable forest management and enhancement of the carbon stocks to mitigate climate change. To reduce emissions from deforestation and forest degradation, there is an urgent need to establish the Measurement, Reporting and Verification (MRV) system. This requires accurate and reliable forest information, which will be useful for estimating forest carbon stocks and carbon accounting (Maniatis, 2010b).

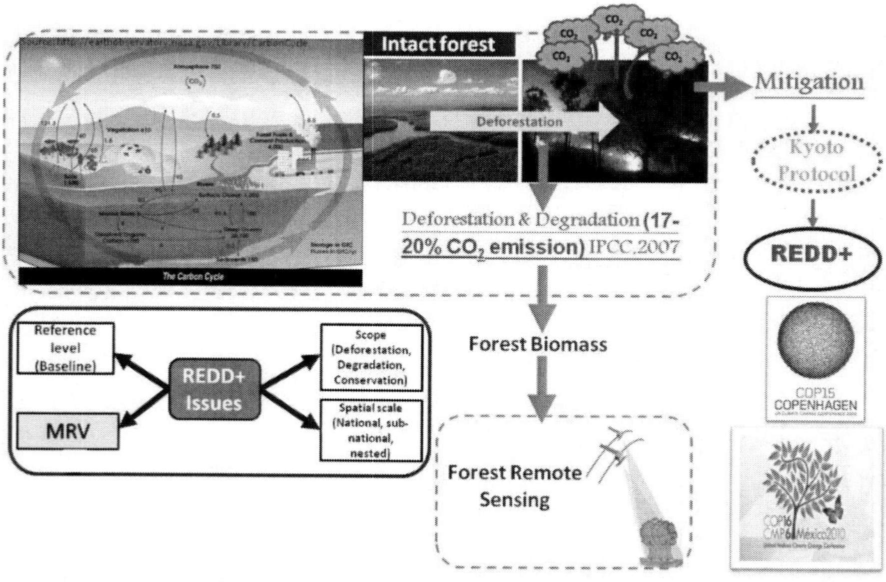

Figure 1.1. Global climate change and mitigation options.

Implementation of REDD+ mechanism will provide financial incentives to the signatory country, which is conserving and enhancing the forest carbon stocks. Therefore, measurement of the forest cover and biomass has become very crucial because such economic assessment requires precise and reliable methods to estimate forest carbon stocks at national level (Gibbs et al., 2007;

Maire et al., 2011; Saatchi et al., 2011). Satellite based remote sensing techniques have played a crucial role in the estimation of forest cover and deforestation monitoring. However, estimation of forest biomass is still under the question because of high uncertainties. In recent studies, it has been proved that remote sensing based forest cover and biomass monitoring is cost and time effective as compared to that of forest inventory data, and it can provide spatial and temporal information of forest cover and biomass distribution (Rosenqvist et al., 2003; DeFries et al., 2007) (Figure 1.1).

1.2. REDD+ An Overview

Reducing emissions from Deforestation and forest Degradation, and the role of conservation, sustainable management of forests and enhancement of forest carbon stocks in developing countries (REDD+) started as a global initiative for climate change mitigation. The basic idea of the REDD+ is to reward individuals, communities, projects and countries that reduce GHG emissions from the forests (REDD, 2010 Source book). The implementation of the REDD+ mechanism require an effective biomass and deforestation monitoring systems that are reproducible, provide consistent results, meet standards for mapping accuracy, and implementable at the national level (Defries, 2007).

1.2.1. What Is REDD+?

The UN-REDD+ programme is United Nations endorsed mechanism to mitigate climate change. It is a programme of three agencies within the UN (i.e., FAO, UNEP and UNDP). UN-REDD+ is an international financial mechanism between developed and developing countries, focusing on reduction of GHG emissions. It was placed on the agenda at COP11 and later on at COP13 where discussions on the methodology and support for its activities were initiated. Recently, COP15 and COP16 have pointed out methodological and policy level strategies respectively. According to UNFCCC, REDD+ can be defined as:

> policy approaches and positive incentives on issues relating to reducing emissions from deforestation and forest degradation in developing countries; and the role of conservation, sustainable

management of forests and enhancement of forest carbon stocks in developing countries.

— UNFCCC Decision 2/CP.13–11

The idea of REDD+ is very simple but implementation of this idea in reality is very challenging. REDD+ is related to social, political and technical issues. For effective REDD+ implementation, emissions reduction must be additional and permanent with no leakage (Angelsen, 2009).

REDD+ carbon-accounting has three main approaches for the geological scale: (i) National level (ii) Sub-National level and (iii) Nested approach. Each approach has some pros and cons. The national approach accounts for national level baseline, carbon accounting, monitoring, and verification system. In the national level approach, carbon credits are owned and managed by the national government. The sub-national approach accounts for sub-national level baseline, carbon accounting, monitoring and verification. In sub-national approach carbon credits are owned and managed by project proponents. A nested approach allows countries to start with a sub-national approach and to scale up to a national approach over time, or to simultaneously account and receive credits at both the sub-national and national levels. The nested approach, therefore, is more flexible and allows more countries to take part in REDD+ (Angelsen, 2009).

1.2.2. Phased Approach

As indicated in the COP16 Decision (Draft Decision -/CP.16 III C Article 73), REDD+ will be implemented in phases to allow countries to participate in the mechanism while taking account of their national circumstances. The countries interested in implementing REDD+ have to follow three phases of REDD+ (a) Readiness phase (b) Demonstration phase and (c) Results based action. Currently most of the countries are in the first phase and UNFCCC is helping to getting ready for REDD+. In the readiness phase (a) countries develop national strategies or action plans, policies, measures, and capacity-building. Following this, in the implementation two more phases will take place: (b) the implementation of national policies and measures and national strategies or action plans that could further involve capacity-building, technology development and transfer and results-based demonstration activities and the final phase is (c) results-based actions that should be fully measured, reported and verified (Draft Decision -CP.16 III C Article73) (Figure 1.2). Table 1.1 shows the various phases of REDD+ mechanism implementation.

Countries willing to participate in REDD+ mitigation mechanism under the UNFCCC, will have to come forward and establish two key points i.e., (i) a Reference Level and/or Reference Emission Level (RL/REL), and (ii) a national forest monitoring system to support the MRV function of the convention (Maniatis, 2010).

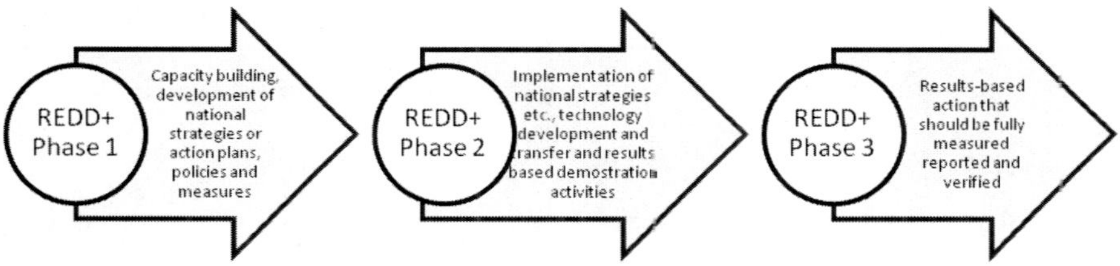

Figure 1.2. Three phased system for REDD+ implementation (Source: Maniatis, 2010).

Table 1.1. Elements of phased approach

No.		Phase 1	Phase 2	Phase 3
1	Scope	RED/REDD/REDD+	REDD/REDD+	REDD+
2	Crediting scale	Sub-national	Nested (both Sub-national and national)	Nested or national approach
3	Performance indicators	-Strategy adopted -Legislative and policy assessment completed -Consultations conducted -Institutions in place	-Policies enacted -Measures enforced -Proxies for forest carbon changes	Quantified forest carbon changes (tCO_2e), compared to a reference level
4	Funding	-Initial support for national strategy development and readiness activities (e.g., FCPF, REDD, bilateral initiatives)	-Funding from bilateral and multilateral sources and COP-mandated funds.	Primarily linked to compliance carbon markets, but might also be via global fund
5	MRV systems	Capacity development	-Capacity development and basic monitoring capacities	Advanced monitoring capacities and setting reference levels

Source: Meridian Institute (2009a, b); Angelsen, 2009.

1.3. OBJECTIVES

The objective of the book is to focus on use of geospatial techniques to monitor forest cover, deforestation and forest biomass, needed for the REDD+ implementation and sustainable forest management planning. Use of advanced remote sensing techniques has potential to monitor effectively. This book is focused on use of Synthetic Aperture Radar (SAR) data to monitor three priority issues that are forest cover, deforestation and forest biomass in Cambodia. Figure 1.3 shows the research framework of the study that is covered in this book mainly focusing on (a) Forest cover mapping (b) Deforestation monitoring (c) Forest biomass estimation using geospatial data.

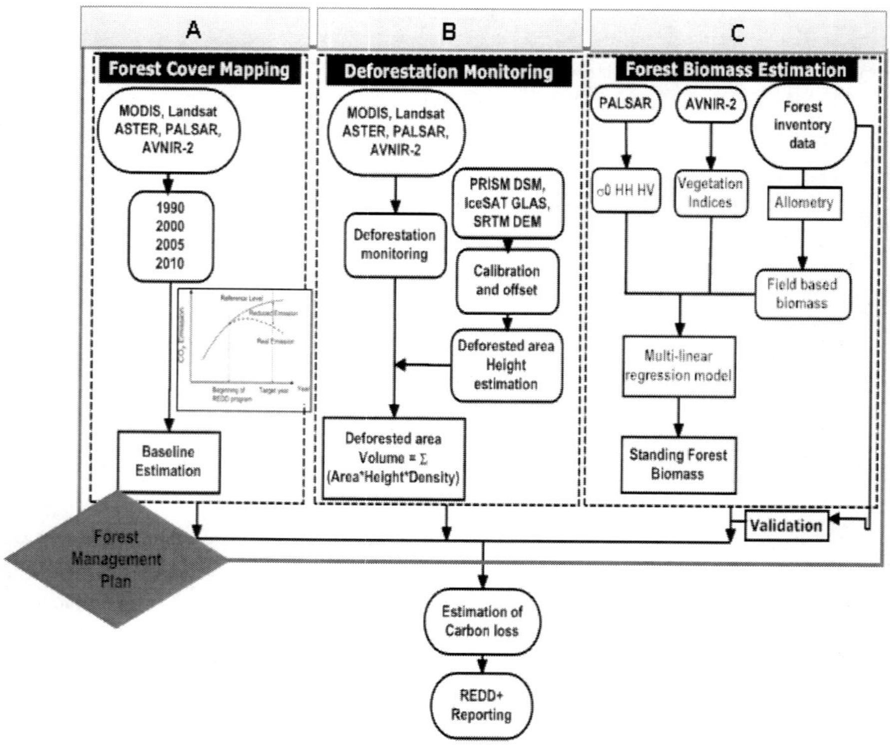

Figure 1.3. Research framework.

1.4. BOOK OUTLINE

This book consists of eight chapters. Chapter 1 provides the overall introduction of the forests and their importance and role of REDD+ mechanism to manage the forest in post Kyoto climate change mitigation regime. Chapter 2 covers the study area (Cambodia) and the forest management plan of Cambodian government. Chapter 3 explains the use of full polarimetric PALSAR data and various polarimetric decomposition model to monitor forest cover. Chapter 4 describes the characterization of the forests and deforestation using full polarimetric PALSAR data based on backscattering (σ^o) and polarimetric parameters of PALSAR. Chapter 5 presents the estimation of the height of cut over forest using various DEMs data. Chapter 6 explains the importance of PALSAR 50m mosaic data for the national level forest biomass estimation. Chapter 7 explains the forest management in Cambodia and use of remote sensing techniques for effective forest management and REDD+ mechanism implementation. Chapter 8 draws the overall conclusions of the book and the recommendations for the future work.

These chapters may be considered as useful scientific and methodological contribution to the existing literature for establishing MRV system, because estimation of carbon loss due to deforestation and forest biomass is critical for MRV establishment to strengthen REDD+ policies implementation. While, Subsidiary Body on Scientific and Technological Advice (SBSTA) is working on MRV establishment but until now there has been no existing MRV system for forest monitoring. This research work will provide motivations for the further work in the field of climate change mitigation. Besides this there are some limitations of this study, which will be considered in the future research work.

Chapter 2

Cambodian Forest

2.1. Geography and Climate

Cambodia is located in Southeast Asia between 10° – 15°N latitude and 102° – 108°E longitude, covering about 181,037 km^2 of area. Cambodia shares its border with Vietnam to the east, Thailand to the west, Lao P.D.R. to the north and Gulf of Thailand to the south. Cambodia is a country with extensive plain lands and Tonle Sap (Great lake), which crosses the country from the north to the south. Topographically Cambodia is divided into two parts: (i) the central low lying or the central plains and (ii) the mountainous ranges. Central plains, consisting mainly of the alluvial plain of Mekong River and the Tonle Sap, cover about three quarters of the country's area. The Cardamom and Elephant Mountain of the southwest, Dangreak Mountain of north and Ratnakiri plateau of east are the sites of densely forested and sparsely populated area. The Mekong River flows through Cambodia with a total length of 486 km. The Tonle Sap is the largest freshwater lake in Southeast Asia with a size of about 2600 km^2 in dry season and increase to 13000 km^2 in rainy season (UNEP, 2009). The Mekong River supplies the water into Tonle Sap during the rainy season and the water flows out from Tonle Sap lake to Mekong River during the dry season. Geologically, Cambodia is covered by alluvium, shale, sandstone and conglomerates on hilly regions (Crocker, 1961).

Cambodia is a tropical country with two distinct seasons: the dry season from November to April and the rainy season from May to October. The mean annual precipitation depends on the region and ranges from 100 to 300 cm (Kiyono et al., 2010). The heaviest rainfall, over 300 cm per year, occurs along the western coastal lowland area. Relative humidity ranges from 65-70% in January and February to 85-90% in August and September. Cambodia's average temperature ranges from 20° to 35°C. April is the warmest month, when the temperature can rise above 38°C and coldest is January with about 22°C (UNEP, 2009).

2.2. Forest Resources

Forests are one of the important natural resources of Cambodia. They provide lots of ecosystem services as well as timber and non-timber forest products. Cambodia's forests are a rich common property resource that plays a crucial role in rural livelihood. Forest resources support subsistence and income generating activities such as small scale timber harvesting, fuel wood collection, resin tapping, and collection of wild fruits, vegetables and medicines. Many rural households view agriculture as their primary employment and the collection of forest resources as a vital secondary and tertiary occupation (UNEP, 2009, Blaser et al., 2011). Figure 2.1 shows the trend of forest cover change in Cambodia. According to forestry administration data, Cambodia was covered by about 73% of forest in 1965, and 59% in 2006 respectively. Cambodia has lost about 14% of forest cover in the last 40 years (Figure 2.1).

Forest types of Cambodia can be broadly divided into (a) evergreen, (b) deciduous and (c) semi-evergreen forests. Evergreen forests are located in areas that receive high rainfall mainly located in the Cardamom Mountains, the northern uplands, and the northern provinces of Kampong thom, Stung Treng and Preah Vihear. Deciduous forests are located in lowland areas of Mondolkiri, Kratie, southern Ratnakiri, Stung Treng, Preah Vihear and Pursat. Flooded forests are located around the Tonle Sap and throughout the Mekong River delta. The main tree species of evergreen forests are Dipterocarpus dyeri, D. corbatus, D. alatus, Anisoptera cochinchinensis, Hopea spp, Roherea vulgaris and Syzygium spp. The main tree species of deciduous forests are Dipterocarpus intricatus, D. obtusifolius and Shorea obtusa. The main tree species of semi-evergreen forests are Dipterocarpus intricatus, D. obtusifolius and Shorea obtusa (Blaser et al., 2011). Table 2.1 shows the percentage of forest cover types in Cambodia according to Forestry Administration (FA), Cambodia 2006 report. This table shows that about 26% of Cambodia is covered by deciduous forests followed by evergreen forests (20%).

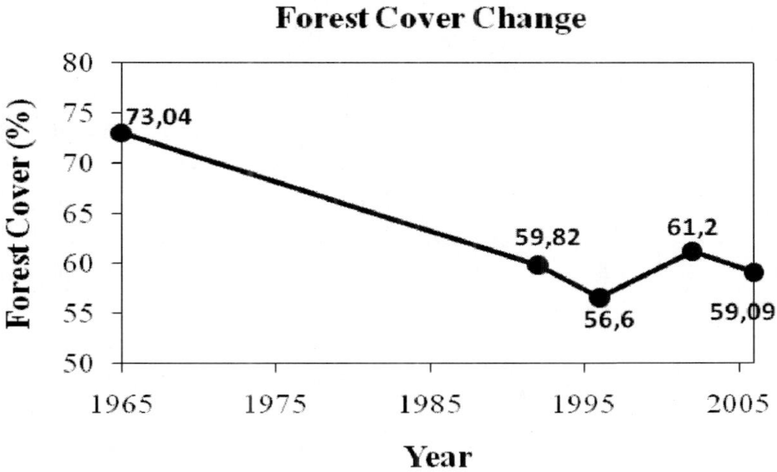

Figure 2.1. Forest cover change in Cambodia.

Table 2.1. Forest Cover in Cambodia, 2006 (FA)

No.	Forest Types	Area (ha)	Percentage (%)
1	Evergreen Forest	3,668,902	20.2
2	Semi-evergreen Forest	1,362,638	7.5
3	Deciduous Forest	4,692,098	25.8
4	Wood Shrub land Dry	37,028	0.2
5	Wood Shrub land Evergreen	96,387	0.5
6	Bamboo	35,802	0.2
7	Other Forest	971,341	5.3
8	Total Forest	10,864,186	59.8
9	Non Forest	7,296,456	40.2
	Total Area	18,160,674	100.0

2.3. Deforestation and Forest Degradation in Cambodia

Forest resources are at great risk due to increasing demand and population growth. Driving factors of deforestation and forest degradation are variable and complex and it can be divided into two types: (a) direct causes and (b) background causes. The direct causes include plantation development, expansion of agriculture and forest fire. Background causes include poverty, unclear land ownership, population growth, economic development etc. The deforestation and forest degradation in an extensive area not only have a negative impact to lower the functionality or value of forests, but also have a variety of impacts on surroundings. It sometimes causes various environmental problems such as soil erosion, landslides, debris flow, which may result in a threat to daily life of local people (JAFTA, 2010). Table 2.2 shows the recent data of Forest Resource Assessment (FRA, 2010). What is alarming here is; Cambodia shows the highest deforestation rate among the Indo-China countries. FRA (2010) estimated that forest cover declined by about 637,000 hectares between 2005 and 2010 and by 2.85 million hectares between 1990 and 2010 (Table 2.2). The Government of Cambodia reported that, in the period 2003–2007, 899,000 hectares of forest were converted legally to agricultural uses and another 224,000 hectares were converted illegally. Estimates of the extent of illegal logging, shifting cultivation, impacts of fire, drought, pests and diseases on the forest estate were unavailable. In the period 2002–06, the most significant loss of forests occurred in the Banteay, Meanchey, Battambong, Siem Reap, Oddar Meanchey and Pailin provinces of Cambodia (Forest Administration, 2010).

Amariei (2004) has reported that the main causes of illegal activities include a faulty legal system, insufficient knowledge and poor management, excessive discretionary powers in the public and private sectors, poor implementation capacity of the public forest administration and enforcement agencies and lack of transparency in decision-making in the forestry sector at all levels. Other contributing factors are the unclear tenure and use rights. To control deforestation and forest degradation in Cambodia, there is an urgent need to control illegal logging as well as make an effective forest management plan.

Table 2.2. Forest cover change in Indo-China countries (FRA, 2010)

Country	FOREST COVER									
	Area				Annual change rate					
	1990	2000	2005	2010	1990-2000		2000-2005		2005-2010	
	1000 ha	1000 ha	1000 ha	1000 ha	1000 ha/yr	%	1000 ha/yr	%	1000 ha/yr	%
Cambodia	12,944	11,546	10,731	10,094	-140	-1.1	-163	-1.4	-127	-1.2
Lao PDR	17,314	16,532	16,142	15,751	-78	-0.5	-78	-0.5	-78	-0.5
Myanmar	39,218	34,868	32,321	31,773	-435	-1.2	-309	-0.9	-310	-0.9
Thailand	19,549	19,004	18,898	18,972	-55	-0.3	-21	-0.1	15	0.08
Viet Nam	9,363	11,725	13,077	13,797	236	2.3	270	2.2	144	1.08

2.4. Forest Concession Management in Cambodia

Economic land concession is a mechanism which provides companies such as business enterprises with land use rights of national land and private land for the purpose of agriculture use and industrial/agricultural development. At present, Cambodia has 85 concession companies with a total land area of 956,690 ha located in 16 provinces (Economic land concession, MAFF, 2010). Table 2.3 shows the economic land concession area in various provinces of Cambodia.

Table 2.3. Number of economic land concession and areas (2006)

Province	no. of companies	no. of concessions	Area (ha)
Stung Treng	10	10	179,899
Kampong Speu	8	9	90,256
Kratie	7	7	64,373
Ratanakiri	5	5	53,747
Kampong Cham	5	5	12,070
Kampong Thom	4	4	35,561
Kon Kong	3	3	79,300
Kampot	3	3	36,200
Siem Reap	3	3	19,235
Mondolkiri	2	2	17,600
Oddar Meanchey	2	2	16,000
Sihanoukville	2	2	12,800
Pursat	2	2	141,963
Kampong Chhnang	1	1	176,065
Battambang	1	1	8,000
Total	58	59	943,069

Ref. Economic land concessions in Cambodia.

The objectives of the Economic Land Concessions are the following:

- To develop intensive agricultural and industrial-agricultural activities that require a high rate and appropriate level of initial capital investment;
- To achieve a specific set of agreements from the investor for developing the land in an appropriate and perpetual manner based on a land use plan for the area;

- To increase employment in rural areas within a framework of intensification and diversification of livelihood opportunities and within a framework of natural resource management based on appropriate ecological system;
- To encourage small as well as large investments in economic land concession projects; and
- To generate state revenues or the provincial or communal revenues through economic land use fees, taxation and related services charges.

2.5. Community Forest Management

Community forestry is broadly an effort to support and empower communities to continue their traditional use of forest resources and encourage sustainable practices (Tyler, 2006). In Cambodia, the first community project was introduced in 1994 in Takeo province and an approach to involve local people in forest conservation was successful (Heng and Shigeru, 2002). After that, there has been growing trend towards engaging local communities in forest management, both in planning and field activities over the past decade (Oberndorf, 2006). In community forestry projects, local people have given the responsibilities of protecting degraded forest lands and planting of trees. In return, they have been granted the permission to use forest resources and to access agricultural land (Heng and Shigeru, 2002). The World Bank report on Cambodia: A vision for forestry sector development (1999) shows 'high biomass stocked forests' are available for commercial production and 'small forest areas with sparse trees' are managed by local communities. More recently community forestry has received considerable attention as a potential alternative to forest concession management. Table 2.4 shows the total community forestry area of Cambodia. According to recent report of FA (2009), a total of 390 community forestry project have been running in 20 provinces of Cambodia, which covers an area of 367,943 ha area. Among 20 provinces, Kampong Thom covers maximum number of community forestry projects.

Measurement of carbon loss due to deforestation and forest degradation using remote sensing techniques has some limitations. Therefore, involvement of local people to monitor forest is potentially cheaper, as noticed in previous studies (Burgess et al. 2010; Phelps et al. 2010; Skutsch 2010). Hence, promotion of community based forest monitoring and conservation could help to overcome the challenges to measure forest biomass (Danielsen et al., 2011).

Table 2.4. Total Community Forestry Areas (FA, 2009)

No.	Provinces	No. of CF	Areas (ha)
1	Banteay Meanchey	09	4,970
2	Battembang	17	5,415
3	Kampong Cham	10	3,480
4	Kampong Chhnang	33	10,908
5	Kampong Speu	22	12,915
6	Kampong Thom	67	66,065
7	Kampot	18	8,221
8	Koh Kong	02	3,638
9	Kratie	13	24,866
10	Mondulkiri	09	12,401
11	Preah Vihear	21	44,950
12	Pursat	52	6,260
13	Rattanakiri	28	37,815
14	Siem Reap	37	18,122
15	Steung Treng	06	14,838
16	Svay Rien	02	510
17	Takeo	13	10,605
18	Oddor Meanchey	12	60,477
19	Keb	02	426
20	Pailin	04	858

2.6. REDD+ IN CAMBODIA

In Cambodia, two REDD+ pilot projects have been running: (a) Community Forestry Carbon Offset Project (CFCOP) in Oddar Meanchey province and (b) the Seima Protection Forest Project (SPF) in Mundulkiri province. These projects have been implemented in community forest areas and protection forests. In both projects, local communities are granted forest access rights. The primary goal of this project is to successfully enhance storage and sequestration of carbon in the natural forests of Cambodia under emerging REDD+ initiative and to assess a climate-related payment mechanism for forest conservation. The secondary goal includes supporting the implementation of the national community forestry program, securing long-term tenure rights for forest-dependent communities, responding to rural livelihood needs, conserving biodiversity, and supporting hydrological regimes (Poffenberger et al., 2009; Donal, 2011).

The Cambodia REDD+ taskforce has developed the REDD+ national readiness roadmap that set out the activities to be implemented over the next few years and to establish a national REDD+ program. The Royal Government of Cambodia (RGC) strongly supports the REDD+ mechanism and believes not only that REDD+ will help to reduce GHG emissions, but also contribute in reducing alleviating poverty, improving forest governance, and enhancing sustainable forest management in the nation (FA, 2008; Donal, 2011). Butler (2006) has estimated the potential earnings from REDD+ about US$ 85 – US$875 million per year for Cambodia. It depends on the extent of forest protected for REDD+ and the market price of carbon credits.

Chapter 3

FOREST COVER MONITORING BASED ON FULL POLARIMETRIC PALSAR DATA

3.1. INTRODUCTION

Mapping of forest area is necessary for sustainable management and utilization of forest resources. Forests play an important role in terrestrial carbon flux and therefore, any significant change in the forest cover from local to global scale will affect the carbon cycle. Forest cover mapping through ground surveys are labor intensive, time consuming, relatively infrequent, very expensive and these maps become easily outdated (Asner, 2009). Development of remote sensing techniques made mapping of forest cover cost and time effective with significant accuracy. In tropical countries, the frequent cloud cover often restrains the acquisition of optical remote sensing data. Radar techniques become a feasible way of acquiring remotely sensed data within a given time frame because they are able to collect data in all weather conditions. Due to this unique feature, radar data have been used extensively in many fields, including land cover mapping (Lu, 2006; Palmann et al., 2008). Polarimetric SAR data play an important role in land cover classification (Zebkar et al., 1991; Cloude and Pottier 1997).

In January 2006, Japan Aerospace Exploration Agency (JAXA) successfully launched Advanced Land Observation Satellite (ALOS) having L-band Synthetic Aperture Radar (SAR) sensor. Phased Array L-band type Synthetic Aperture Radar (PALSAR) has different mode of acquisitions: single polarization, dual polarization, quad polarization. PALSAR full polarimetric data are advantageous to characterize various targets on the basis of physical, dielectric and geometric properties (Ainsworth et al., 2009). The ability of PALSAR data to retrieve information about physical properties of the surface, has led to a wide range of applications, such as monitoring of crops, hydrology, soil moisture, disaster, ocean and forests (Lee and Pottier, 2009). In this chapter, PALSAR data have been used for forest cover classification. Co-polarized (HH, VV) and cross-polarized (HV) signals are used to determine surface scattering, double-bounce scattering and volume scattering respectively (Rignot et al., 1995). Various decomposition techniques are available for the land cover classification however, in this chapter, eigen vector-based (Cloude-Pottier) and model-based (Freeman-Durden three component and Yamaguchi four component) techniques have been compared. Supervised and unsupervised classification approach using SAR data have been done by various researchers (Lee et al., 1999; Lee et al., 2004; Fang et al., 2006; Alberga, 2007; Ainsworth, et al., 2009). This chapter demonstrates the supervised classification of PALSAR full polarimetric data for forest cover classification. The performance of supervised classification depends on the applications and the availability of ground truth data (Ince, 2010).

Cloude-Pottier decomposition for polarimetric classification, is based on eigen values and eigen vectors to estimate polarimetric scattering of various targets which are derived from the coherency matrix of full polarimetric PALSAR data (Cloude and Pottier, 1996; 1997). Entropy (H), alpha (α) and anisotropy (A) are eigen values based polarimetric decomposition, used to understand the polarimetric response of SAR data to study purity of target, mean scattering, and number of scattering involved for each target (Lee et al., 2004; Patel et al., 2009). Freeman-Durden three component decomposition (1998) approach is the first model based incoherent decomposition. It has been used to model the polarimetric coherence matrix as the contribution of three scattering mechanisms: surface scattering, double-bounce scattering and volume scattering (An, et al., 2011). The four component scattering decomposition technique proposed by Yamaguchi et al. (2006) which is an extension of Freeman-Durden three component decomposition has also been used in this study.

The important role of forest in the global carbon cycle has been pointed out in previous studies (Schlamadinger and Marland, 1996; Brown, 2002, Defries et al., 2007). Therefore, it becomes necessary to monitor forest covers for effective implementation of the REDD+ mechanism. Most of the previous studies (Table. 3.1) show the polarimetric application of SAR data using various decomposition theories. The selection of the best SAR based decomposition classification for forest cover monitoring is highly desirable in order to optimally select the appropriate methodology. Hence, a quantitative comparison between the classification results of various decomposition methods has been made in order to understand the potential of the full polarimetric PALSAR data to generate forest cover map.

Table 3.1. Previous studies related to polarimetric decomposition

No.	Authors	Polarimetric Parameters Used
1	Cloude and Pottier 1997	Entropy and Alpha based classification
2	Lee et al., 1999	H/A/Alpha Wishart classification
3	Ferro-Famil et al., 2001	H/A/Alpha Wishart classification
4	Fang and Wen, 2005	Intensity, Alpha, Entropy
5	Ouarzeddine et al., 2005	Entropy, Alpha, Anisotropy
6	Fang et al., 2006	Entropy, Alpha, Span
7	Alberga, 2007	Entropy, Alpha, Anisotropy based land cover classification
8	Patel et al., 2009	Entropy, Alpha and Anisotropy
9	Wang et al., 2009	Krogagger decomposition with T matrix
10	Ince, 2010	Dynamic clustering of entropy, alpha and Anisotropy
11	An et al., 2011	Four component decomposition with deorientation
12	Singh et al., 2011	Four component scattering power decomposition

3.2. Study Area

Cambodia has the highest rate of deforestation among Indo-China countries (FRA, 2010). Between 2000 and 2005 the country has lost 29% of its primary forests, while overall deforestation was 218,800 ha of forest per year (FAO, 2005). Logging activities combined with rapid development, population growth, urbanization and agricultural expansion have been the primary reasons for Cambodia's forest loss (Gaughan et al., 2009; UNEP, 2009). Cambodia has signed REDD+ mechanism in 2009 and study of forest cover is necessary for the REDD+ implementation. The target area is situated in parts of Stung Treng and Kratie province of northern Cambodia and covers an area of about 1852 Km^2 (Figure 3.1). The area was selected based on the analysis of Landsat data. This area was severely affected by logging activities and covers both evergreen and deciduous types of forests of Cambodia that covers major percentage of Cambodian forests (Table 2.1). Field visits were made in December, 2009 and January, 2011 for the collection of ground truth data. The GPS positions and GPS photos of various land cover types were collected to use as training data for supervised classification of the remote sensing data.

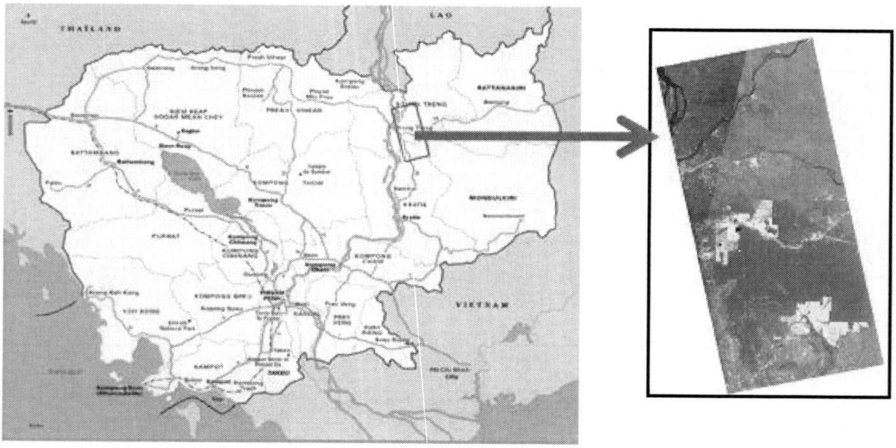

Figure 3.1. Location of the study area.

3.2.1. Land Cover Map of the Study Area

Land cover classification is useful for effective management of land resources and remote sensing technique plays essential role for land use land cover classification (Garcia-Mora et al., 2011). Forests are considered to have high biodiversity and commercial importance and play a significant role in social, economic and cultural development of rural people (Kummer and Turner, 1994). Land cover map (Figure 3.2) was generated using the maximum likelihood classification (MLC) technique. AVNIR-2 data acquired on March, 2010 was used to generate land cover map of the area supported with ground truth data that was collected during the field work in December, 2009. The classes were selected based on the major types of land cover classes by analyzing existing land cover map of Cambodia. The study area was classified into seven land cover classes viz. evergreen forest, deciduous forest, sparsely deciduous forest, wood and shrub land, deforested area, agricultural land and water. It was found that most of the area is covered by evergreen forest followed by deciduous forest, sparsely deciduous forest and other classes. Table 3.2 shows the percentage of different land cover classes of the area.

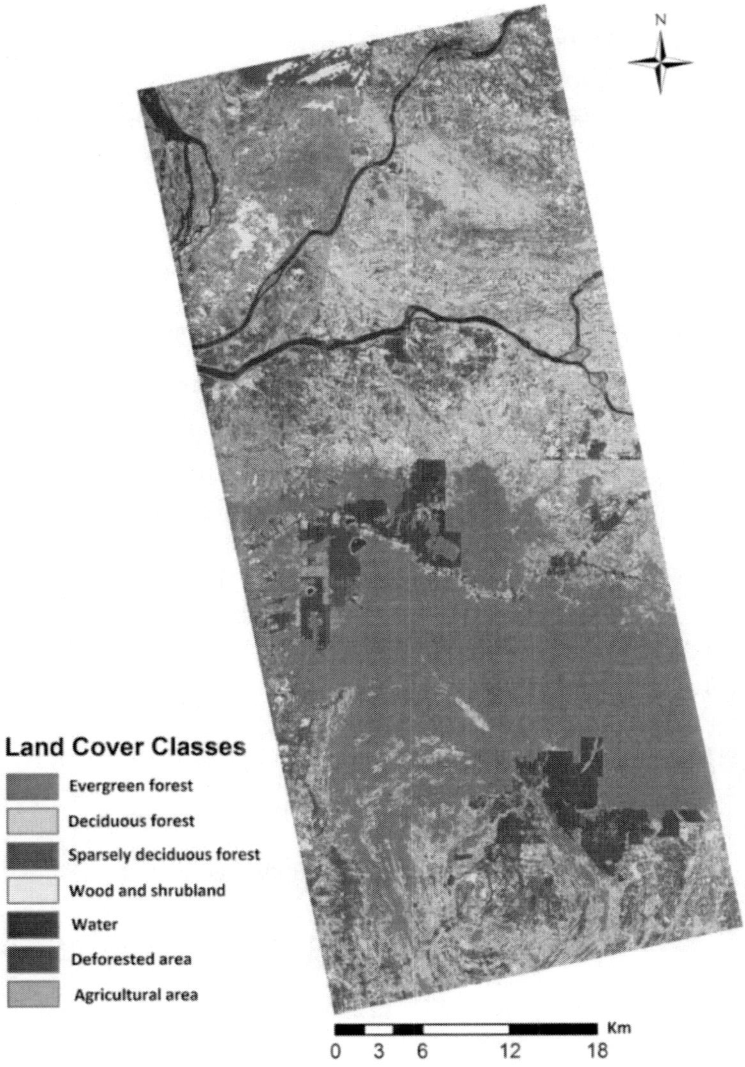

Figure 3.2. Land cover of the study area.

3.3. Methodology

PALSAR 1.1 level full polarimetric, single look complex (SLC) data acquired on April 06, 2009 with 21.5^0 look angle was used as the primary data. PolSARpro4.2 software was used for the data processing. First, the data were multi-looked one time in range and six times in azimuth and converted from slant to ground range, so that the spatial resolution would be reduced to nearly 23m x 21m per pixel (range x azimuth). Then the data were processed with the refined Lee filter with window size 5x5 to reduce the speckle noise (Lee et al., 2008, 2009). The detailed methodology is shown in the flow chart (Figure 3.3).

3.3.1. Polarimetric Decompositions

Polarimetric decomposition provides information about scattering mechanisms or polarimetric properties of the targets. The decomposition method depends on the target types and is divided into two major categories: (a) coherent decomposition (pure targets) and (b) incoherent decomposition (distributed targets). Most of the objects in the study area are natural vegetation, therefore incoherent target decomposition models were used. The incoherent target decomposition models such as Cloude and Pottier (1997), Freeman-Durden three component (1998) and Yamaguchi four component scattering model (2006) were used in this study. These models can be summarized as following:

The coherency matrix is given as

$$[T] = k.k^{*T} \tag{1}$$

where, ^{*T} indicates complex conjugate and transpose, and *k* indicates the target vector.

Target vector is obtained by vectorization V(.) of the scattering matrix S in each pixel:

$$k = V(s) = (1/2)Trace\{S[\psi]\} \tag{2}$$

Ψ is a set of 2 by 2 complex basis matrices.

PALSAR data having monostatic mode, so target vector based on Pauli basis matrix is defined as

$$k = (1/\sqrt{2})\begin{bmatrix} S_{hh} & + & S_{vv} \\ S_{hh} & - & S_{vv} \\ 2S_{hv} & & \end{bmatrix} \tag{3}$$

where, S_{hh}, S_{vv} and S_{hv} are the elements of scattering matrix S

Using equations (1) and (3), coherency matrix [T] can be written as

$$[T]=(1/2)\begin{bmatrix} \langle |S_{hh}+S_{vv}|^2 \rangle & \langle (S_{hh}+S_{vv})(S_{hh}-S_{vv})^* \rangle & 2\langle (S_{hh}+S_{vv})S_{hv}^* \rangle \\ \langle (S_{hh}-S_{vv})(S_{hh}+S_{vv})^* \rangle & \langle |S_{hh}-S_{vv}|^2 \rangle & 2\langle (S_{hh}-S_{vv})S_{hv}^* \rangle \\ 2\langle S_{hv}(S_{hh}+S_{vv}^*) \rangle & 2\langle S_{hv}(S_{hh}-S_{vv})^* \rangle & 4\langle |S_{hv}|^2 \rangle \end{bmatrix} \tag{4}$$

The incoherent decomposition theorems can be expressed as

$$[T]=q_1[T_1]+q_2[T_2]+q_3[T_3]+......+q_k[T_k] \tag{5}$$

This is very general expression of incoherent decomposition where, (q_1, q_2, q_3..... q_k) are coefficient to be determined and (T_1, T_2, T_3..... T_k) expansion coherency matrix.

3.3.2. Cloude-Pottier Decomposition

Cloude and Pottier (1997) proposed a method of extraction of mean diffusion based on eigen values/eigen vectors decomposition of coherency matrix. Based on this idea, they defined entropy (H), alpha angle (α) and anisotropy (A). H, α, A are polarimetric features, give an information about scattering properties from the targets. Entropy indicates the statistical disorder (randomness) of the scatterers. Zero (0) value of H indicates a single scattering mechanism, whereas 1 indicates a random mixture of scattering mechanism (no dominant scatterers). H can be expressed as:

$$H = \sum_{i=1}^{3} -P_i \log_3 P_i \qquad (6)$$

$$0 \leq P_i = \frac{\lambda_i}{\sum_{i=1}^{3} \lambda_i} \leq 1 \qquad (7)$$

where P_i is pseudo-probabilities, which can be obtained from the eigen values λ_i represents relative intensity of the $_i$-th scattering process.

The alpha angle (α) is estimated from α_i angle of each of the eigen vectors as follows

$$\overline{\alpha} = \sum_{i=1}^{3} P_i \alpha_i \qquad (8)$$

Alpha angle is based upon the eigen vectors and is a number indicative of the average or dominant scattering mechanism within a specific target. It varies between 0^0 to 90^0. The value of alpha close to 0^0 represents odd-bounce scattering from flat surfaces, close to 45^0 represents dipole or volume scattering, while value of alpha close to 90^0 represents a double bounce scattering.

Anisotropy (A) is a complementary parameter for Entopy (H). Anisotropy measure relative importance of the second and third eigen values of eigen decomposition. It has less information than H/α and it is useful when H>0.7 (Pottier, 1998; Fang et al., 2006; Trisasongko, 2010).

$$A = \frac{\lambda_2 - \lambda_3}{\lambda_2 + \lambda_3} \qquad (9)$$

where, λ_2 and λ_3 are smaller eigen values.

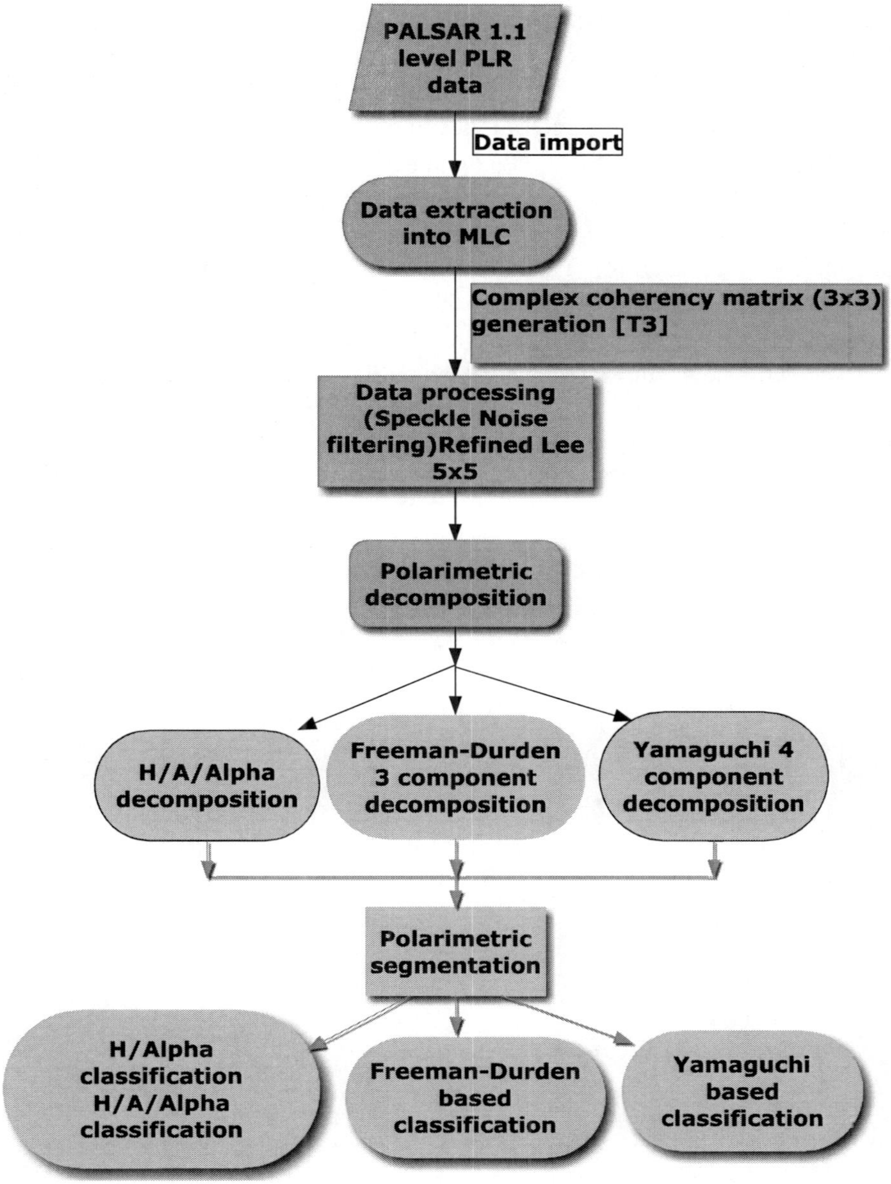

Figure 3.3. Flow chart of the methodology adopted for polarimetric classification.

3.3.3. Target Decomposition Theorem of Freeman and Durden (1998)

The three component scattering decomposition is a technique used for the fitting of a physically based three-component scattering mechanism model to the polarimetric SAR data. Freeman-Durden decomposition is used to derive meaningful information about scatterers characteristics (i.e., volume, odd and double bounce scattering). This approach can be used to determine the first order dominant scattering mechanism that gives rise to observed backscatter in polarimetric SAR image. According to three components scattering decomposition techniques, the coherency matrix can be decomposed into

$$[T] = f_s[T_s] + f_d[T_d] + f_v[T_v] \tag{10}$$

Surface scattering is the first component of three scattering decomposition generated by Bragg surface scatterer

$$[T] = f_s \begin{bmatrix} \langle |\eta+1|^2 \rangle & \langle (\eta+1)(\eta-1)^* \rangle & 0 \\ \langle (\eta+1)^*(\eta-1) \rangle & \langle |\eta+1|^2 \rangle & 0 \\ 0 & 0 & 0 \end{bmatrix} \tag{11}$$

In equation (10) and (11)

$$f_s = |R_{vv}|^2 \text{ and } \eta = \frac{R_{hh}}{R_{vv}} \tag{12}$$

$$R_{hh} = \frac{\cos\theta - \sqrt{\varepsilon_s - \sin^2\theta}}{\cos\theta + \sqrt{\varepsilon_s - \sin^2\theta}} \tag{13}$$

$$R_{vv} = (\varepsilon_s - 1)\frac{\sin^2\theta - \varepsilon_s(1+\sin^2\theta)}{\left[\varepsilon_s\cos\theta + \sqrt{\varepsilon_s - \sin^2\theta}\right]^2} \tag{14}$$

Double bounce scattering is modeled by scattering from a dihedral corner reflector, such as ground-tree trunk backscatter, where the reflector surfaces can be made of different dielectric materials. The coherency matrix for double bounce scattering is represented by

$$[T_d] = f_d \begin{bmatrix} \langle |\chi-1|^2 \rangle & \langle (\chi-1)(\chi+1)^* \rangle & 0 \\ \langle (\chi-1)^*(\chi+1) \rangle & \langle |\chi+1|^2 \rangle & 0 \\ 0 & 0 & 0 \end{bmatrix} \tag{15}$$

Where,

$$f_d = |R_{Tvv}R_{Gvv}|^2 \text{ and } \chi = \frac{R_{hh}}{R_{vv}} \tag{16}$$

where, R_T and R_G are reflection coefficient of vertical trunk surface and Fresnel reflection coefficient of horizontal ground surface.

The volume scattering component is a third scattering component of the decomposition model and can be represented by

$$[T_v] = \frac{f_v}{3}\begin{bmatrix} 4 & 0 & 0 \\ 0 & 2 & 0 \\ 0 & 0 & 2 \end{bmatrix} \tag{17}$$

3.3.4. Target Decomposition Theorem of Yamaguchi (2006)

Based on the Freeman-Durden three component scattering model, Yamaguchi et al. (2006) proposed a four component scattering decomposition technique by introducing helical scattering. This helical scattering power term, corresponds to $\langle S_{HH}S^*_{HV}\rangle \neq 0$ and $\langle S_{HV}S^*_{VV}\rangle \neq 0$ appears in heterogeneous areas (complicated shape targets or man-made structures) whereas disappears for almost all natural distributed scattering (Yamaguchi et al., 2005). The coherency matrix can be decomposed into

$$[T]=f_s[T_s]+f_d[T_d]+f_v[T_v]+f_c[T_c] \tag{18}$$

where f_s, f_d, f_v and f_c are coefficient to be determined. $[T_s]$, $[T_d]$, $[T_v]$ and $[T_c]$ are expansion coherency matrix corresponding to the surface, double-bounce, volume and helical scattering, respectively.

The single bounce scattering model is represented by surface scattering properties from rough surfaces in which the cross-polarized component is negligible. The expansion coherency matrix is expressed as:

$$[T_s]=f_s\begin{bmatrix} 1 & \eta^* & 0 \\ \eta & |\eta|^2 & 0 \\ 0 & 0 & 0 \end{bmatrix} \tag{19}$$

The expansion coherency matrix for double bounce scattering is expressed as:

$$[T_d]=f_d\begin{bmatrix} |\chi|^2 & \chi & 0 \\ \chi^* & 1 & 0 \\ 0 & 0 & 0 \end{bmatrix} \tag{20}$$

The helical scattering power is equivalent to circular polarization power. This expression becomes visible in urban and mountainous areas. The helical scattering expansion matrix, which takes into account of non-reflection symmetry condition, is expressed as:

$$[T_c]=f_c\begin{bmatrix}0 & 0 & 0\\ 0 & 1 & \pm j\\ 0 & \pm j & 1\end{bmatrix} \tag{21}$$

Most of the area in the present study is covered by forest so the helical scattering is not the main component except mountainous area. Hence, emphasis was given to the volume scattering from the forested area.

For the volume scattering model, one of the following matrices can be selected according to the magnitude balance of $|S_{hh}|^2$ and $|S_{vv}|^2$. These equations represent the different distribution of vegetation to discriminate different types of vegetation (Yamaguchi et al., 2005).

When $10\log(|S_{vv}|^2/|S_{hh}|^2) \geq 2$ dB

$$[T_v]=\frac{1}{30}\begin{bmatrix}15 & -5 & 0\\ -5 & 7 & 0\\ 0 & 0 & 8\end{bmatrix} \tag{22}$$

When $10\log(|S_{vv}|^2/|S_{hh}|^2) < 2$ dB

$$[T_v]=\frac{1}{4}\begin{bmatrix}2 & 0 & 0\\ 0 & 1 & 0\\ 0 & 0 & 1\end{bmatrix} \tag{23}$$

When $10\log(|S_{vv}|^2/|S_{hh}|^2) \leq -2$ dB

$$[T_v]=\frac{1}{30}\begin{bmatrix}15 & 5 & 0\\ 5 & 7 & 0\\ 0 & 0 & 8\end{bmatrix} \tag{24}$$

where f_s, f_d, f_v and f_c are surface, double, volume and helical scattering coefficient respectively.

3.3.5. Classification Technique

H/A/Alpha Cloude-Pottier decomposition, Freeman-Durden three component and Yamaguchi four component scattering approaches have been attempted for the classification of PALSAR image. Supervised classification was applied on Cloude-Pottier decomposition, Freeman-Durden three component decomposition, and Yamaguchi four component decomposition. For supervised classification, Maximum likelihood classification (MLC) has been used, which is a parametric classifier that assumes normal or near normal spectral distribution for each feature of interest (Lu et al., 2004). This classifier evaluates variance and covariance when classifying an unknown pixel. Confusion matrix was used to evaluate the performance of a classifier (Kohavi and Provost, 1998). Confusion matrix is a matrix plot between the predicted and defined classes. Kappa coefficient was also used to assess the accuracy of the classification (Congalton R. G., 1991).

3.4. Results and Discussion

The polarimetric decomposition techniques were applied on the PALSAR data. The α, H and A of the PALSAR data have been generated from eigen values based target decomposition, shown in Figure 3.4 a, b and c respectively. The value of alpha (Figure 3.4a) represents the spatial variation. In Figure (3.4a), about 60-70% of the area has shown the values of alpha between 40^0 to 50^0 because of the volume or multiple scattering which is a characteristic of forests. The low value of α (10^0 to 20^0) represents surface scattering, and is a characteristic of plane surfaces like water body or deforested area or agricultural land. As have been stated, entropy indicates statistical disorder or randomness in the targets. Higher values of H are visible in (Figure 3.4b) which means the backscattered signals are much depolarized. This is a characteristic feature of the evergreen forest. The value of H in evergreen forest is higher than deciduous forest (Figure 3.4b). This indicates more randomness in evergreen forest as compared to deciduous forest. In case of evergreen forests high randomness in scattering is because of the mixture of scattering from different components of trees (trunk, branches, crown, leaves etc.) with multi-story canopy structure. The value of H in the deforested area is lower than those surrounding forests. Therefore, decrease in entropy is determined as an identification of deforestation and forest degradation and is confirmed using the ground truth data. The low value of entropy and alpha in the deforested area is because of the surface scattering which causes a reflection of signal far from the sensor. The anisotropy (Figure 3.4c) shows most of the area having a high value of entropy which corresponds to secondary and tertiary scattering. The area with low value of entropy represents dominant first scattering and low secondary scattering.

The H/A/Alpha based R:G:B colour composite image is presented in the figure 3.5a. In this image clearly shows the dense forest and deforested area. Agricultural area shows misclassification with deforested area because both classes show low value of H and alpha. However, agricultural area with crops can be easily identified because of increase in value of H and alpha. It gives an idea about the H/A/Alpha behavior of various targets.

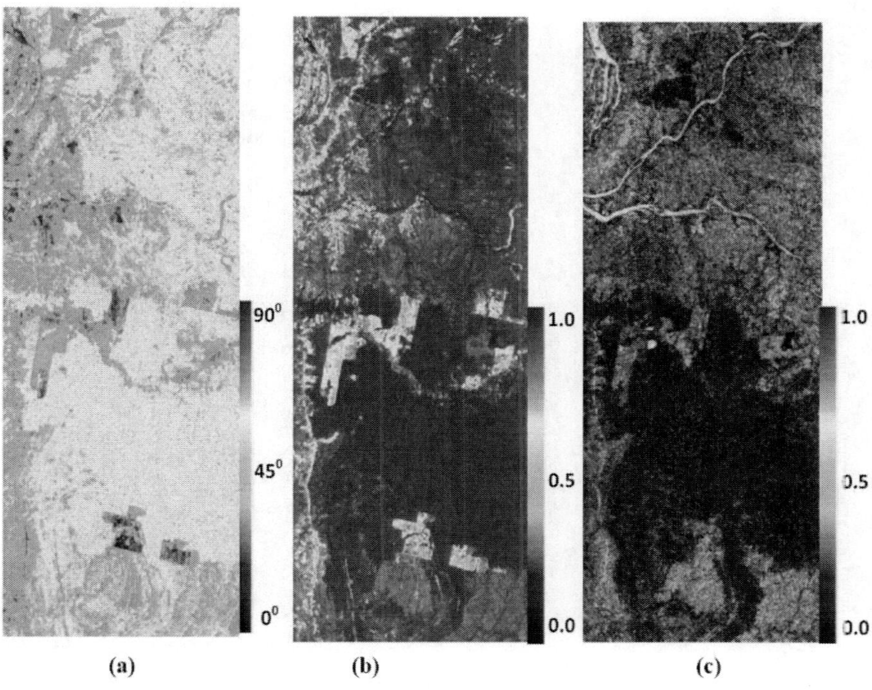

Figure 3.4. (a) Alpha, (b) Entropy and (c) Anisotropy images of PALSAR data.

R:G:B color composite of Freeman-Durden three component decomposition (double bounce assigned red color, volume scattering assigned green color and surface scattering assigned blue color) and Yamaguchi four component decomposition images are shown in figure 3.6a and 3.6b respectively. It gives an approximate percentage of each of the surface, volume and double bounce scattering mechanisms present in the area. From figure 3.6a and 3.6b, it becomes clear that volume scattering is dominant in the Freeman-Durden three component decomposition image than the Yamaguchi four component decomposition image. Most of the evergreen forest shows green color because of higher value of volume backscattering as compared to deciduous forest. In dry season, sparse deciduous forest of the study area shows cyan color because of higher surface backscattering than volume scattering. Deforested area has high surface scattering and can be clearly distinguished from the surrounding forested area (Figure 3.6a and 3.6b).

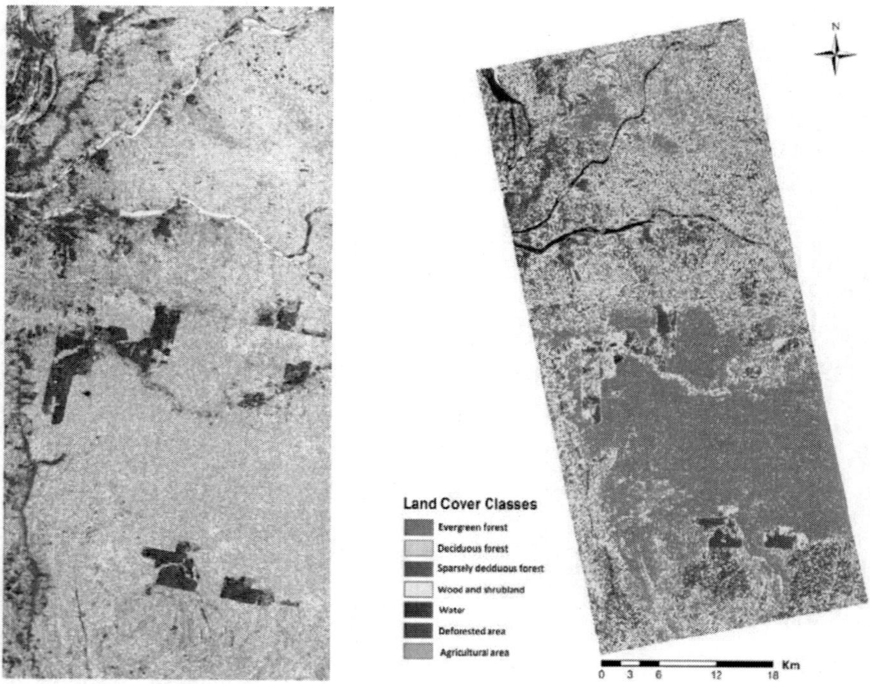

(a) (b)

Figure 3.5. (a) H/A/Alpha (H-Green; A-Red; Alpha-Blue) colour image (b) H/A/Alpha based land use map (MLC).

MLC technique has been applied on figure 3.5a, 3.6a and 3.6b with the same training samples based on field information. Classified PALSAR images using Cloude-Pottier H/A/Alpha decomposition, Freeman-Durden three component and Yamaguchi four component scattering model have been presented in figure 3.5b, 3.8a and 3.8b respectively. PALSAR data have been classified into seven major classes. Table 3.2 illustrates the land cover percentage obtained from the Cloude-Pottier H/A/Alpha decomposition, Freeman-Durden three component and Yamaguchi four component based classification taking AVNIR-2 as a reference.

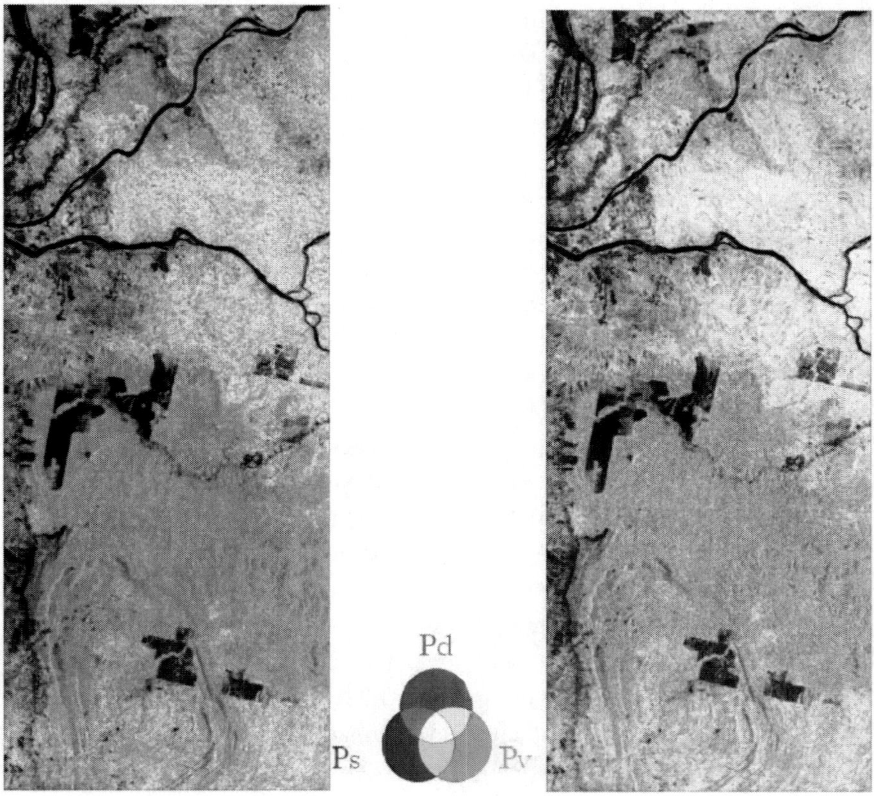

(a) (b)

Figure 3.6. (a) Freeman-Durden three component based image of PALSAR (b) Yamaguchi four component based image of PALSAR (R:Double bounce scattering, G:Volume scattering, B:Surface scattering) (Avtar et al., 2011a).

As given in Table 3.2, the areas of different land cover classes are different for each classification. The area of evergreen forest is over estimated by about 4.7%, 3.7% and 13.9% in Cloude-Pottier classification, Yamaguchi four component and Freeman-Durden three component classification respectively. Freeman-Durden three component classification has shcwn overestimation (13.9%) in evergreen forest may be because of the over estimation problem in volume scattering component. The area of deciduous forest is 7% underestimated in Freeman-Durden classification and 4.8% overestimated in Cloude-Pottier classification, whereas Yamaguchi classification shows the same area (21.1%). Underestimation in deciduous forest, by Freeman-Durden three component classification probably because some of the deciduous forests have shown same volume scattering as evergreen forests did. This can be confirmed based on the field pictures (figure 3.7) because deciduous forests were covered by tall grass, which causes volume scattering. Cloude-Pottier based classification shows 8.3% underestimation in sparsely deciduous forests, which might be because of low value of alpha in sparsely deciduous forest, hence represents surface scattering behaviour. Deforested area, water body and agricultural land may have been somewhat misclassification because flat surfaces have less backscattering in L-band and cannot be distinguished easily. If agricultural land has some crops then it may be easy to distinguish from the flat surface because crop causes volume scattering.

The most common tool to access the accuracy of classification is the confusion matrix. Table 3.3, 3.4 and 3.5 illustrate confusion matrix of Cloude-Pottier MLC, Freeman-Durden three component MLC and Yamaguchi four component MLC based on the predicted and defined pixels respectively. It shows high agreement between the observed estimate and the field measurement. Therefore, it appears that field measurement is necessary to obtain precise measurement of land cover types. The evergreen and sparsely deciduous forests have shown the matrix with 100% producer's accuracy for Freeman-Durden three component and Yamaguchi four component based classification, whereas other types of land cover have shown some variability in the matrix. Overall, Yamaguchi four components based MLC produced highest accuracy with accuracy of 94.1% and 0.93 kappa coefficient than the 90.3% with 0.88 kappa coefficient by Freeman-Durden classification and 89.7% with 0.88 kappa coefficient by Cloude-Pottier classification (Table 3.6). The low accuracy of Cloude-Pottier MLC might be because of the eigen values based decomposition, where the information associated with eigen vectors is not utilized but still it gives good results.

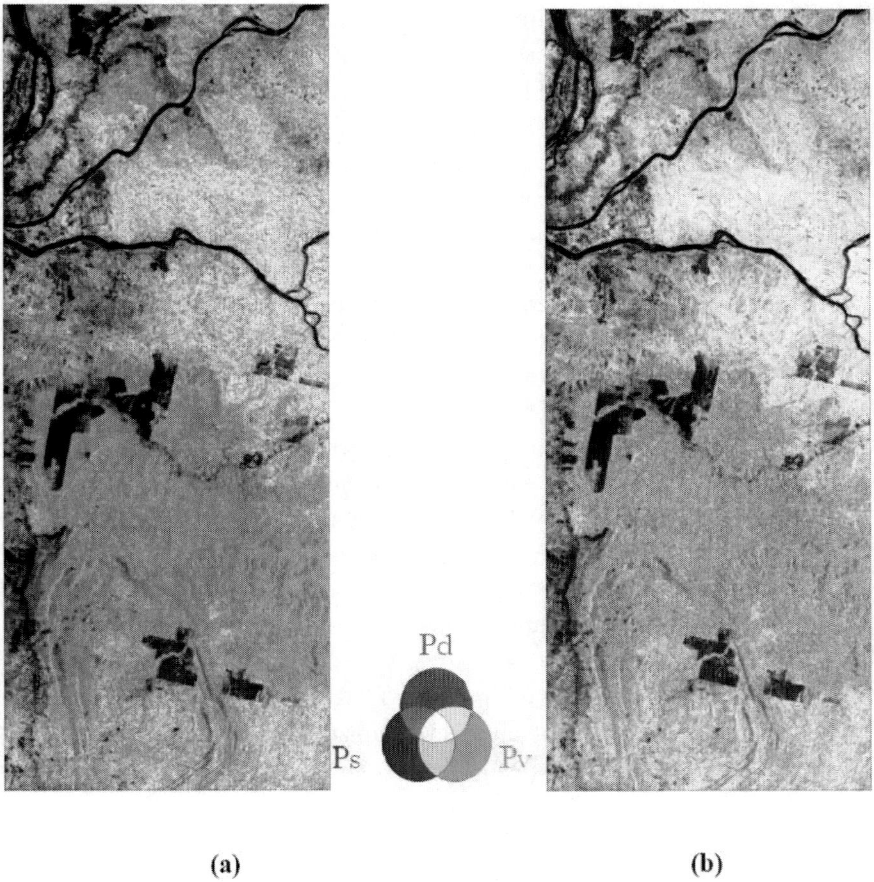

Table 3.2. Percentage of different types of land cover

Land cover classes	AVNIR-2 based classes		Clode-Pottier based classes		Freeman-Durden three component based classes		Yamaguchi four component based classes	
	Area (Km^2)	% of land cover	Area (Km^2)	% of land cover	Area (Km^2)	% of land cover	Area (Km^2)	% of land cover
Evergreen forest	550.1	29.7%	637.4	34.40%	808.4	43.6%	617.9	33.4%
Deciduous forest	388.1	21.0%	477.9	25.80%	264.5	14.3%	391.1	21.1%
Sparsely deciduous forest	387.8	20.9%	232.6	12.60%	309.6	16.7%	456.5	24.6%
Wood and shrub land	245.3	13.2%	261.9	14.14%	233.8	12.6%	176.7	9.5%
Water body	38.4	2.1%	62.8	3.39%	66.1	3.6%	26.4	1.4%
Deforested area	161.0	8.7%	95.1	5.13%	56.5	3.1%	55.3	3.0%
Agricultural land	81.4	4.4%	84.4	4.56%	113.2	6.1%	128.2	6.9%
Total	1852.1	100%	1852.1	100%	1852.1	100%	1852.1	100%

Figure 3.7. Picture showing deciduous forest covered by grass (>50cm height).

Table 3.3. Confusion matrix of Cloude-Potteir H/A/Alpha based MLC

	Evergreen forest	Deciduous forest	Water body	Sparsely Deciduous forest	Wood and shrub land	Deforested area	Agricultural land	Total	Users Accuracy
Evergreen forest	60	3	0	0	0	0	0	**63**	**95.2%**
Deciduous forest	1	43	4	1	0	0	0	**49**	**87.7%**
Water body	0	2	61	0	0	0	0	**63**	**96.8%**
Sparsely Deciduous forest	0	0	5	49	2	0	1	**57**	**85.9%**
Wood and shrub land	0	1	0	1	29	1	2	**34**	**85.3%**
Deforested area	0	0	2	0	0	49	6	**57**	**85.9%**
Agricultural land	0	0	0	0	1	2	16	**19**	**84.2%**
Total	**61**	**49**	**72**	**51**	**32**	**52**	**25**	**342**	
Producer's Accuracy	**98.4%**	**87.8%**	**84.7%**	**96.1%**	**90.6%**	**94.2%**	**64%**		

Table 3.4. Confusion matrix of Freeman-Durden three component based MLC

	Evergreen forest	**Deciduous forest**	**Water body**	**Sparsely Deciduous forest**	**Wood and shrub land**	**Deforested area**	**Agricultural land**	**Total**	**Users Accuracy**
Evergreen forest	61	4	0	0	1	0	0	**66**	**92.4%**
Deciduous forest	0	44	0	0	3	0	0	**47**	**93.6%**
Water body	0	0	66	0	0	7	9	**82**	**80.5%**
Sparsely Deciduous forest	0	0	0	53	0	0	0	**53**	**100%**
Wood and shrub land	0	1	0	0	29	0	0	**30**	**96.6%**
Deforested area	0	0	6	0	0	44	2	**52**	**84.6%**
Agricultural land	0	0	0	0	0	0	12	**12**	**100%**
Total	**61**	**49**	**72**	**53**	**33**	**51**	**23**	**342**	
Producer's Accuracy	**100%**	**90%**	**92%**	**100%**	**87%**	**86%**	**52%**		

Table 3.5. Confusion matrix of Yamaguchi four component based MLC

	Evergreen forest	**Deciduous forest**	**Water body**	**Sparsely Deciduous forest**	**Wood and shrub land**	**Deforested area**	**Agricultural land**	**Total**	**Users Accuracy**
Evergreen forest	61	0	0	0	0	0	0	**61**	**100%**
Deciduous forest	0	49	0	0	0	0	0	**49**	**100%**
Water body	0	0	64	0	0	0	0	**64**	**100%**
Sparsely Deciduous forest	0	0	0	53	0	0	0	**53**	**100%**
Wood and shrub land	0	0	0	0	33	0	0	**33**	**100%**
Deforested area	0	0	8	0	0	46	7	**61**	**75.4%**
Agricultural land	0	0	0	0	0	5	16	**21**	**76.2%**
Total	**61**	**49**	72	**53**	**33**	**51**	**23**	**342**	
Producer's Accuracy	**100%**	**100%**	**89%**	**100%**	**100%**	**90%**	**70%**		

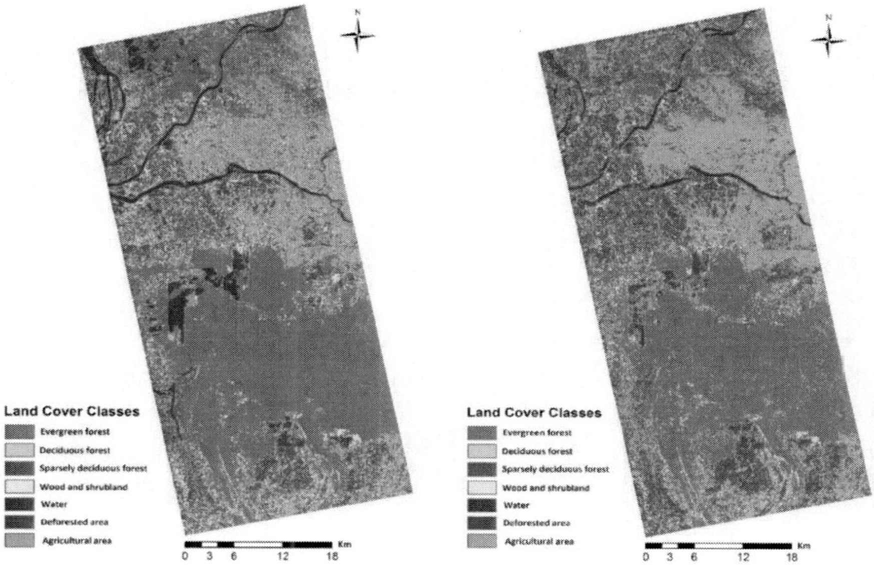

(a) (b)

Figure 3.8. (a) Freeman-Durden three components based Land use map (b) Yamaguchi four component based Land use map (Maximum Likelihood Classification) (Avtar et al., 2011a).

Table 3.6. Accuracy assessment of MLC

	Overall accuracy	**Kappa coefficient**
Cloude-Pottier based classification	89.7%	0.88
Freeman-Durden based classification	90.3%	0.88
Yamaguchi based classification	94.1%	0.93

CONCLUSION

This study demonstrates the potential of using full polarimetric L-band PALSAR data for land cover classification. Cloude-Pottier H/A/Alpha, Freeman-Durden three components and Yamaguchi four component decomposition were used to study land cover in Cambodia. The evergreen and sparsely deciduous forests have shown the confusion matrix with high accuracy for both Freeman-Durden three components and Yamaguchi four component based classification. Freeman-Durden three component classification shows overestimation in volume scattering in evergreen forest. It is interesting that overall accuracy of Yamaguchi four components based MLC (94.1%) is more as compared to the Freeman three component based classification (90.3%) and Cloude-Pottier based classification (89.7%). Results show that use of full polarimetric data improves the detection of structural differences between forest canopies, thus helping in better forest mapping and management practices. The use of PALSAR data for forests cover classification can overcome problems of clouds and hazes present in optical data as well as identification of deciduous forests during dry season. The new generation of full polarimetric SAR technology could help to study forest parameters for better implementation of REDD+ mechanism. The applicability of this methodology is limited to flat areas because of topographic effects of SAR data.

Chapter 4

FORESTS AND DEFORESTATION CHARACTERIZATION USING PALSAR DATA

4.1. INTRODUCTION

Forests have a vital socio-economic and environmental importance and play an important role in maintaining ecological balance and homeostatic in the environment (Lackey, 1998). The Tropical forests provide a significant benefit to humans at the local and global scales (Gardner et al., 2009).

For the implementation of REDD+ mechanism, an accurate assessment of (a) place of deforestation with forest-types, (b) aerial extent (ha), (c) biomass lost (in percentage) (d) carbon contents in each forest-type (metric tons of carbon per hectare), and (e) the process of forest loss are needed (Ramankutty et al., 2007).

The traditional field-based assessment of the forest distribution is a tedious, expensive and time consuming process; especially, making an assessment at a large scale in remote areas (Knuth et al., 2009). Therefore, remote sensing is a useful tool to monitor the distribution of forest at various spatial and temporal scales, when combined with the ground truth data. It is also an advantageous tool due to its repetitive measurements at large scale synoptic view (Lu et al., 2006). Currently, most of the operational systems used for deforestation monitoring are based on the optical sensors (Gibbs et al., 2007b). However, launching of the Synthetic Aperture Radar (SAR) based satellites (e.g., JERS, PALSAR, RADARSAT-1/2, ERS-1/2, ENVISAT, TerraSAR-X, TanDEM-X etc.) have opened a wide spectrum for regular and cloud-free observations in tropical regions (Kuntz, 2010). SAR based observations can provide information about dielectric properties, roughness, orientation etc. of the objects. Most of the previous studies (Le Toan et al., 1992; Beaudoin et al., 1994; Dobson et al., 1995, Watanabe et al., 2006, Mitchard et al., 2009) have been done using SAR backscattering coefficient (σ^o) as a main parameter for forest cover and deforestation monitoring. This chapter attempts to examine various other polarimetric parameters to characterize forests and deforestation in Cambodia. In this study, relative polarimetric descriptors *viz.* RVI, PH, H, α and A has been used to characterize forest-types and deforestation, as these polarimetric descriptors have minor effects from the topography, incidence angle and other calibration errors (Kim and Vanzyl, 2001).

This chapter is an attempt to develop a better understanding to discriminate forest types and deforestation based on polarimetric behavior of PALSAR so that effective implementation of REDD+ policies can be undertaken.

4.2. Methodology

4.2.1. Satellite Data

Two scenes of full polarimetric (HH, HV, VH, VV) 1.1 level PALSAR data were acquired on May 17, 2007 and April 6, 2009 with 21.5^0 look angle. Data from the Landsat Thematic Mapper (TM5) and AVNIR-2 were also acquired in June 2006 and March 2010, respectively, and used for the selection of forest cover types. High resolution (2.5m) PRISM data acquired on November, 2009 with the nadir and backward look angle were also used for the selection of the area of interest (AOI). Along with satellite data field based experience were also used for selection of AOIs.

4.2.2. Field Data

The primary data, such as forest types, species, diameter at breast height (DBH), tree height and tree density are essential for carbon stock estimation. These forest inventory data were collected on the basis of stratified random sampling during December 2009 and January 2011. The homogeneous site has been selected on the basis of analysis of above mentioned satellite data. Four plots of evergreen forest, four plots of deciduous and three plots of sparsely deciduous forest were selected for forest inventory data. The 30x60m and 30x30m plots size have been selected on the basis of position and homogenity of the site. The GPS positions and GPS photos of enroute forest cover types were also collected for further selection of plots for quantitative analysis. The biomass of the sampling plots has been calculated using Kiyono et al., (2010) allometric equation. Table 4.1 summarized the details of forest inventory data.

Table 4.1. Forest inventory data of different forests types

No.	Forest Types	Mean Height (m)	Mean DBH (cm)	Tree density (no./ha)	Plot Size (ha)	Biomass (t/ha)
1	Evergreen_1	20.32	21.96	544	0.18	254
2	Evergreen_2	18.3	25.39	338	0.18	480
3	Evergreen_3	17	19.7	532	0.09	170.5
4	Evergreen_4	15.4	19.3	577	0.09	166.2
5	Deciduous_1	7.49	20.94	311	0.18	102
6	Deciduous_2	13.5	19.7	577	0.09	149
7	Deciduous_3	12	21.6	288	0.09	141
8	Deciduous_4	11	21.4	400	0.09	121.5
9	Sparsely Deciduous_1	10.5	20.6	222	0.09	58.7
10	Sparsely Deciduous_2	11.6	26.1	188	0.09	86
11	Sparsely Deciduous_3	10.34	25.53	166	0.18	73

4.2.3. Data Processing

For the processing of PALSAR data- NEST, PolSARpro and ENVI software were used. The PALSAR images were calibrated radiometrically through converting the digital number to the normalized radar cross section in a decibel. The formula is given below (Shimada et al., 2009).

$$\sigma^o = 10 \times \log_{10}(I^2 + Q^2) + CF - 32.0 \quad (1)$$

Where, CF is calibration factor (CF = -86 dB) and I and Q are the real and imaginary part of the complex SAR image pixel values, respectively. PALSAR 1.1 level full polarimetric, single look complex (SLC) data were multi-looked at one time in range and six times in azimuth direction, converted from the slant to ground range so that the spatial resolution is reduced to nearly 21m x 23m (azimuth x range). Later, the data was filtered using Frost filter with window size of 5x5 to reduce speckle effects. All images were geo-referenced within pixel accuracy.

The PALSAR data were analyzed to represent different types of image regions, representing the different types of forest cover classes. An area of interest (AOI) was selected in homogeneous area on the basis of land cover map (Figure 4.2). The 40 AOIs were selected to analyze different types of polarimetric behaviour of PALSAR data. The selected AOIs were 12, 10, 9, and 9 in evergreen, deciduous, sparsely deciduous forests, and deforested areas, respectively. The different polarimetric behaviours of the PALSAR data were analyzed for each selected forest cover classes (AOI). Figure 4.1 represents PALSAR data with the location of the sampling plots (AOI) of different types of forests. Figure 4.2 shows the methodology adopted in this chpater. The backscattering (σ^o) properties depend on target characteristics. It shows surface, volume and double bounce backscattering behaviours of the targets.

4.2.4. Polarimetric Decomposition Parameters

Polarimetric decomposition provides information to understand the physical characteristics of various scatterers. It is based on the eigen values and eigen vectors (Cloude and Pottier, 1997). The entropy (H), alpha (α) angle and anisotropy (A) are polarimetric features, which gives the information about scattering properties from the targets. H, α and A were used to characterize forests and deforestation.

4.2.5. Radar Vegetation Index

Radar Vegetation Index (RVI) is used for analyzing the scattering from the vegetated area showing high value of volume scattering targets (leaves, branches etc.) (Qi et al., 2010). Woody vegetation, having high cross polarization components shows high values of RVI. RVI is derived from the radar backscattering coefficient (σ^o) of HH, HV and VV polarizations (Kim and Van Zyl, 2001).

$$RVI = 8\sigma^o HV / \sigma^o HH + \sigma^o VV + 2\sigma^o HV \tag{2}$$

4.2.6. Pedestal Height

Pedestal height (PH) is used for measuring randomness in the scattering process. PH is simply a ratio of minimum eigenvalue (λ_3) to the maximum eigenvalue (λ_1). These eigenvalues correspond to the minimum and maximum powers achievable by optimizing over all antenna transmit and receive polarization (Lee and Pottier, 2009). PH value varies from 0 to 1.

$$PH = \min(\lambda_1, \lambda_2, \lambda_3) / \max(\lambda 1, \lambda 2, \lambda_3) = \lambda_3 / \lambda_1 \quad (3)$$

With $\lambda_3 \leq \lambda_2 \leq \lambda_1$ $0 \leq PH \leq 1$

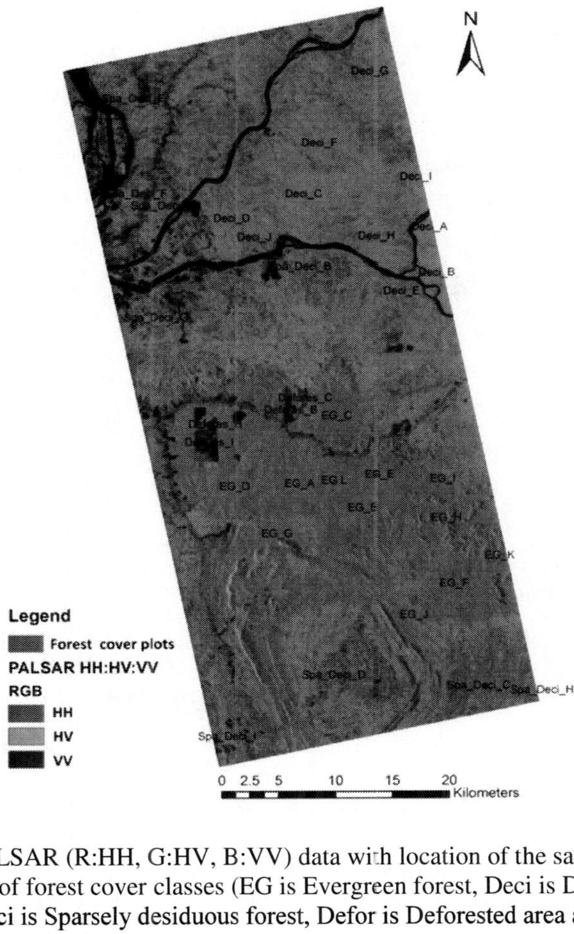

Figure 4.1. PALSAR (R:HH, G:HV, B:VV) data with location of the sampling plots of different types of forest cover classes (EG is Evergreen forest, Deci is Deciduous forest, Spa_Deci is Sparsely desiduous forest, Defor is Deforested area and a,b,c,d..... is plot ID of particular forest class).

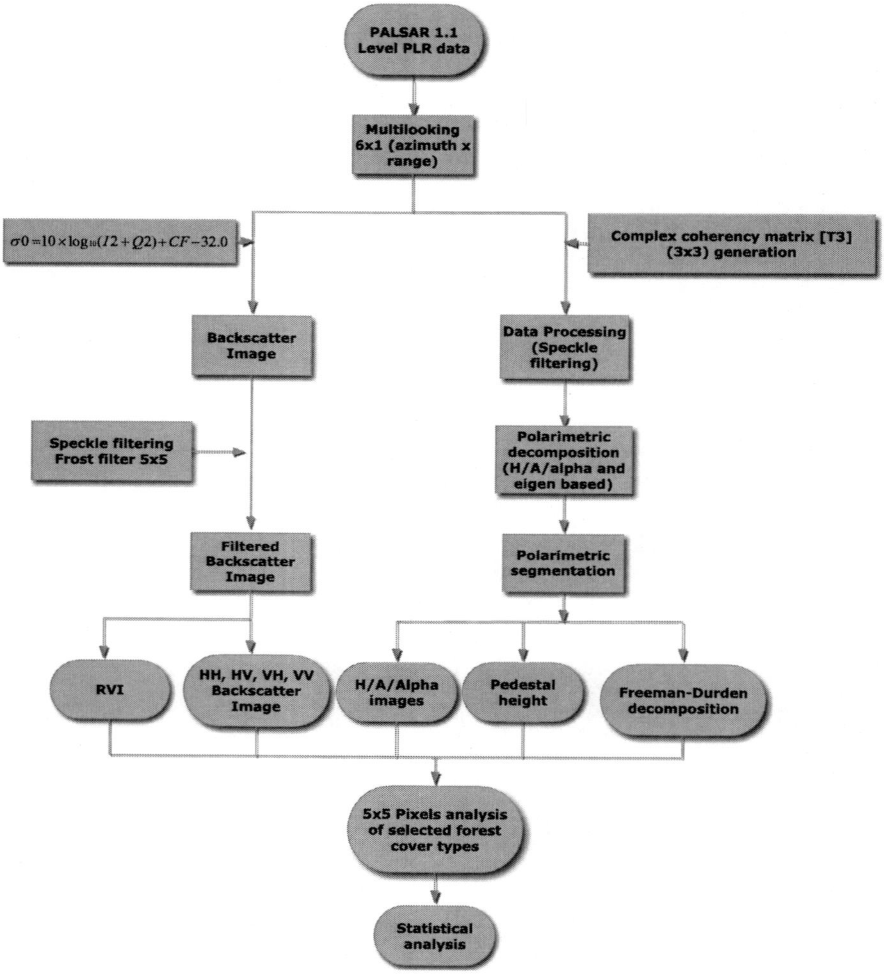

Figure 4.2. Flow chart of the methodology (Avtar et al., 2011b).

4.2.7. Three-Component and Four-Component Based Target Decomposition

Three-component scattering decomposition is a technique used for the fitting of a physically based three-component scattering mechanism model to polarimetric SAR data. Freeman and Durden decomposition (1998) is used to derive meaningful information about the scatterers' characteristics (i. e. surface, volume, and double bounce scattering). These surface, volume and double bounce scattering mechanisms are useful for discrimination of various forest types and deforested area (Lee and Pottier, 2009). Yamaguchi four-component based target decomposition was also used to identify the forests and non-forested area.

4.3. Results and Discussion

Figure 4.3 shows the variations of mean backscattering coefficients (σ^o) of HH, HV, VH, and VV with different types of forest classes. The value of σ^o depends on the polarization types. HH, HV and VV polarization represent surface scattering, canopy scattering and ground-trunk scattering respectively in forest areas (Toan et al., 1992). These three scattering mechanisms are dependent on the polarimetry of forest-types. The L-band backscattering coefficients have shown a good correlation with biophysical parameters (biomass and height) of the forest. Similar features have been observed at various locations for different types of forests (Le Toan et al., 1992; Beaudoin et al., 1994; Dobson et al., 1995, Watanabe et al., 2006, Mitchard et al., 2009). Table 4.2 illustrates the statistical summary of the observed polarimetric parameters for different types of forests. The mean σ^o HV observed for evergreen, deciduous, sparsely deciduous forests and deforested area were -12.96, -12.99, -14.29, and -22.25 (dB), respectively. The σ^o HV of evergreen forest was comparatively higher than those observed for the other forest types (Table 4.2). This observed feature from evergreen forest was due to the fact that evergreen forest with multiple canopy enhances the volume scattering. The mean value of σ^o HV backscattering for deciduous forest was higher compared to the sparsely deciduous forest and deforested area. This observation shows a saturation stage of the σ^o HV, because deciduous forest has biomass greater than 100 tons/ha, which is a saturation limit of the σ^o HV for PALSAR. Watanabe et al., (2006) has also found saturation of PALSAR signal at 100 tons/ha. The observed mean σ^o HH shows a higher value for deciduous and sparsely deciduous forests, whereas lower for evergreen (Table 4.2). This is due to the higher penetration of L-band (low incidence angle 23.1^0) which caused surface scattering. The value of mean σ^o HV is very low for deforested area due to the absence of volume scattering. The mean σ^o HH for some deforested plot has shown a higher value in the recent cut over forest but it decreases over a period of time. This high value of σ^o HH in recent deforested area might be because of the presence of wooden logs, which increases surface roughness (Isoguchi et al., 2009). Therefore, in case of deforested area, the σ° HV shows a sudden change as compared to σ° HH.

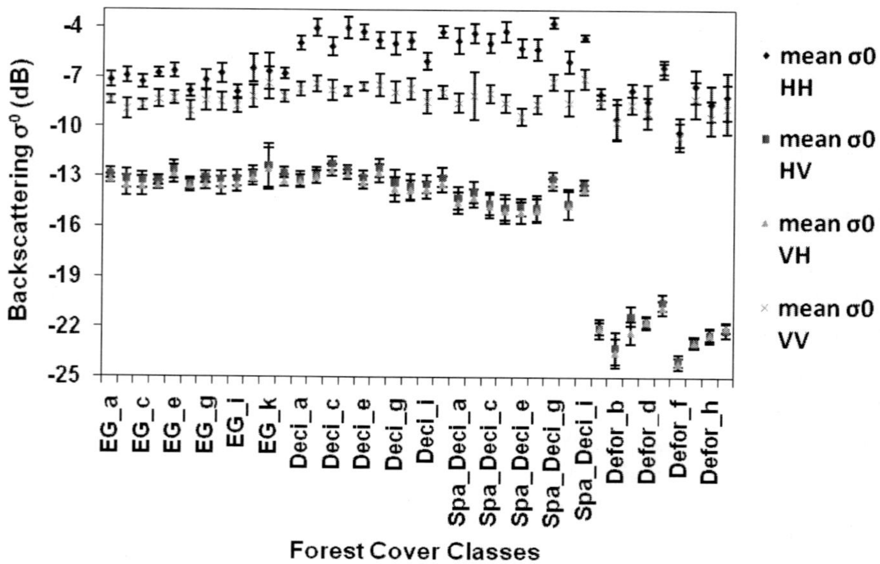

Figure 4.3. Graph represents the relation between PALSAR backscattering versus forest cover classes (EG is Evergreen forest, Deci is Deciduous forest, Spa_Deci is Sparsely desiduous forest, Defor is Deforested area and a,b,c,d….. is plot ID of particular forest class).

Figure 4.4 shows the variations of HH/HV polarization ratio with different types of forest classes, and the observed values of HH/HV are illustrated in Table 4.2. The mean value of the HH/HV polarization ratio was lower for evergreen forest (5.95 dB) than those observed for deciduous (8.25 dB) and sparsely deciduous forests (9.48 dB). On the other hand, the HH/HV polarization ratio observed for the deforested area is highest (13.96 dB). The variations of HH/VV co-polarization ratio is high for deciduous (3.05 dB) and sparsely deciduous forests (3.43 dB), whereas lower for evergreen forests (1.39 dB). In case of the deforested area, the value of HH/VV co-polarization ratio is slightly low (0.57 dB). Hence, the HH/HV (cross-polarization) polarization ratio is better than HH/VV (co-polarization ratio) to identify the type of forests and deforestation because of 2-4 dB difference in backscattering.

Table 4.2. Statistical summary of polarimetric parameters for different types of forest

	Evergreen forest (mean ± SD)	Deciduous forest (mean ± SD)	Sparsely Deciduous forest (mean ± SD)	Deforested area (mean ± SD)
σ0 HH (dB)	-7.04 ± 0.46	-4.73 ± 0.61	-4.81 ± 0.68	-8.29 ± 1.13
σ0 HV (dB)	-12.96 ± 0.34	-12.99 ± 0.44	-14.29 ± 0.63	-22.25 ± 1.03
σ0 VH (dB)	-13.30 ± 0.34	-13.27 ± 0.44	-14.52 ± 0.59	-22.42 ± 0.99
σ0 VV (dB)	-8.43 ± 0.44	-7.78 ± 0.29	-8.23 ± 0.71	-8.86 ± 1.1
σ0 HH/HV (dB)	5.95 ± 0.37	8.25 ± 0.65	9.48 ± 0.56	13.96 ± 0.6
σ0 HH/VV (dB)	1.39 ± 0.36	3.05 ± 0.46	3.43 ± 0.65	0.57 ± 0.28
Alpha (Degree)	46.25 ± 1.12	46.44 ± 1.3	43.03 ± 1.65	11.17 ± 1.8
Entropy	0.93 ± 0.01	0.85 ± 0.03	0.80 ± 0.03	0.36 ± 0.04
Anisotropy	0.12 ± 0.03	0.23 ± 0.05	0.26 ± 0.04	0.25 ± 0.05
RVI	0.91 ± 0.04	0.67 ± 0.06	0.54 ± 0.04	0.17 ± 0.02
Pedestal Height	0.41 ± 0.03	0.26 ± 0.04	0.20 ± 0.02	0.04 ± 0.01
Freeman-Durden_surface	0.03 ± 0.02	0.12 ± 0.06	0.22 ± 0.05	0.25 ± 0.08
Freeman-Durden_volume	0.43 ± 0.03	0.42 ± 0.05	0.31 ± 0.05	0.04 ± 0.01
Freeman-Durden_double bounce	0.01 ± 0.01	0.11 ± 0.06	0.05 ± 0.02	0.01 ± 0.002

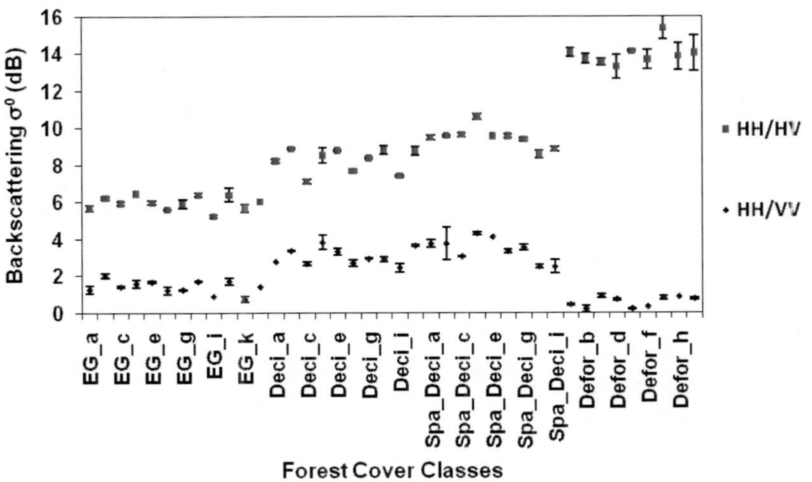

Figure 4.4. Variations in the cross-polar and co-polar ratios with forest cover classes.

Figure 4.5 (a) show the false colour composite of the Landsat TM5 data of June 2006, while figure 4.5 (b, c) shows R:HH, G:HV, and B:VV colour composite of the PALSAR backscatter images observed during May 2007 and April 2009. From the comparison of the images in figure 4.5 (a,b,c), it can be estimated that the area of cut over forest of about 20.13 Km^2 and 43.32 Km^2, occurred from the year 2006 to 2007, and 2007 to 2009, respectively. This deforested area was calculated based on digitization of deforested area interpreted on PALSAR data. The entropy (H), alpha angle (α), anisotropy (A) images from PALSAR data have been generated using eigen value-based target decomposition (Figure 4.6 (a,b,c)). The H, α and A based R:G:B colour composite is shown in figure 4.7. The value of α ranges from 40^0 to 50^0 for evergreen, deciduous, and sparsely deciduous forests. On the other hand, it decreases to 10^0 to 15^0 for deforested area due to surface scattering (Figure 4.8, Table 4.2). The mean value of the entropy for evergreen forest is higher (0.93) than those shown by deciduous (0.85) and sparsely deciduous (0.80) forest (Figure 4.9). Forests show a high value of entropy because of distributed targets, which causes depolarization effects due to volume scattering from the vegetation. The value of entropy becomes less for deforested area because most of the deforested area shows the pure target properties (Guerra et al., 2008). Therefore, decreasing the entropy can be regarded as deforestation. However, it should be confirmed by ground truth observation. The value of anisotropy shows small changes for deforested area (Figure 4.9).

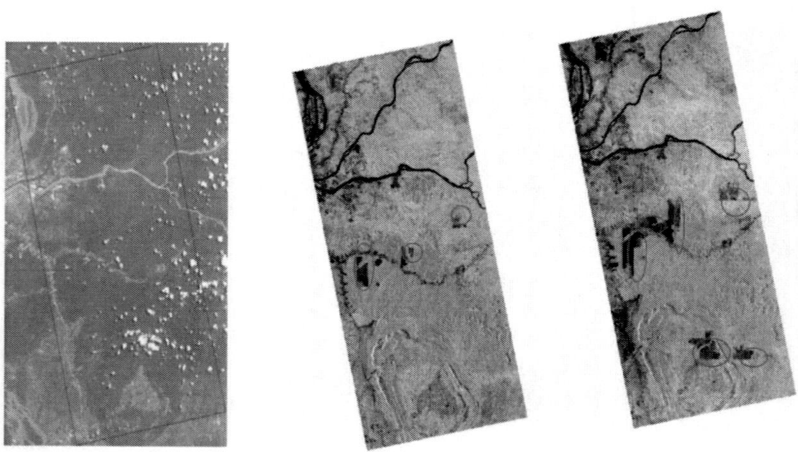

a b c

Figure 4.5. (a) Landsat TM image of 08/06/2006 (b) PALSAR (R:HH, G:HV, B:VV) image of 17/05/2007, and (c) PALSAR (R:HH, G:HV, B:VV) image of 06/04/2009.

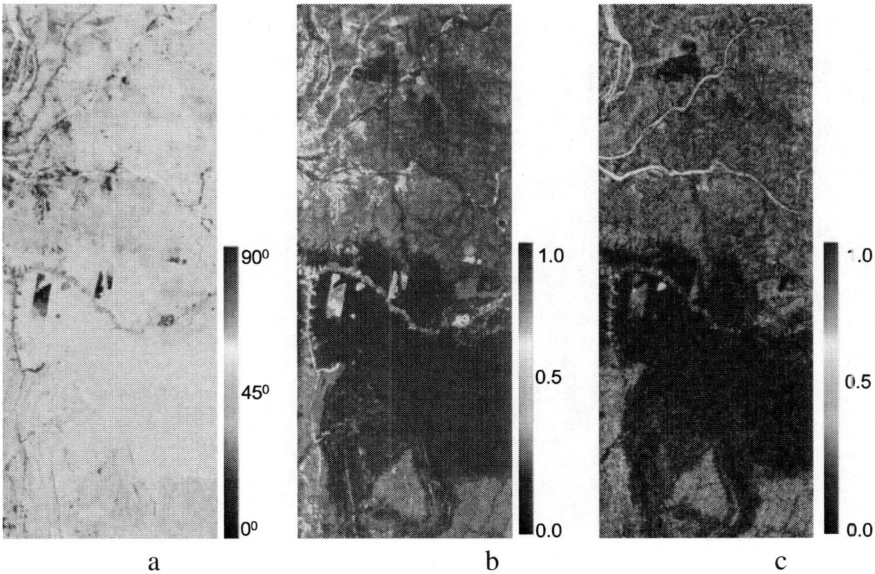

Figure 4.6. a) Alpha, b) Entropy and c) Anisotropy images from PALSAR data.

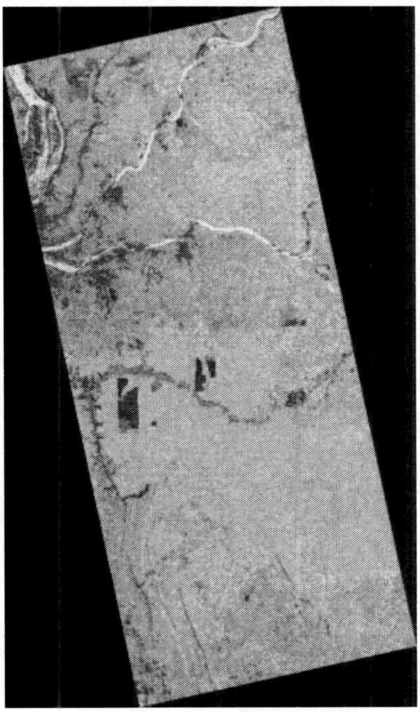

Figure 4.7. H/A/Alpha (H-Green; A-Red; Alpha-Blue) colour image.

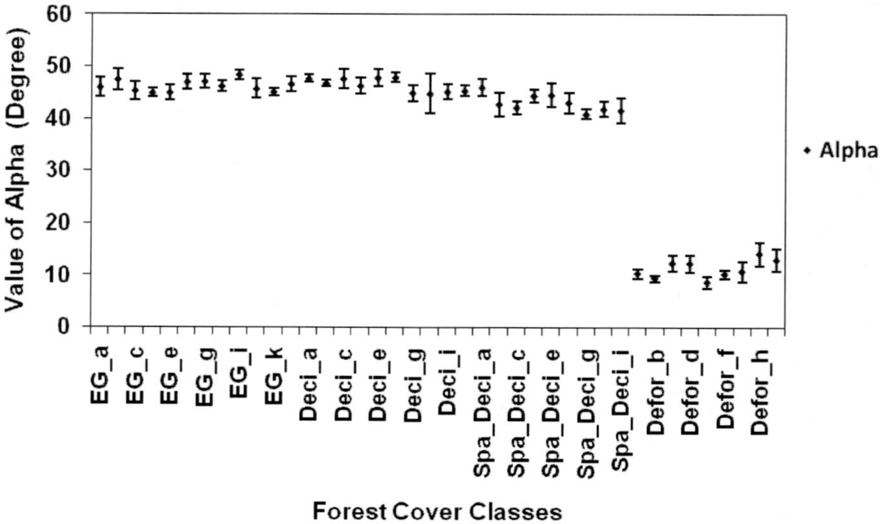

Figure 4.8 Variations of the Alpha (α) angle with forest cover classes.

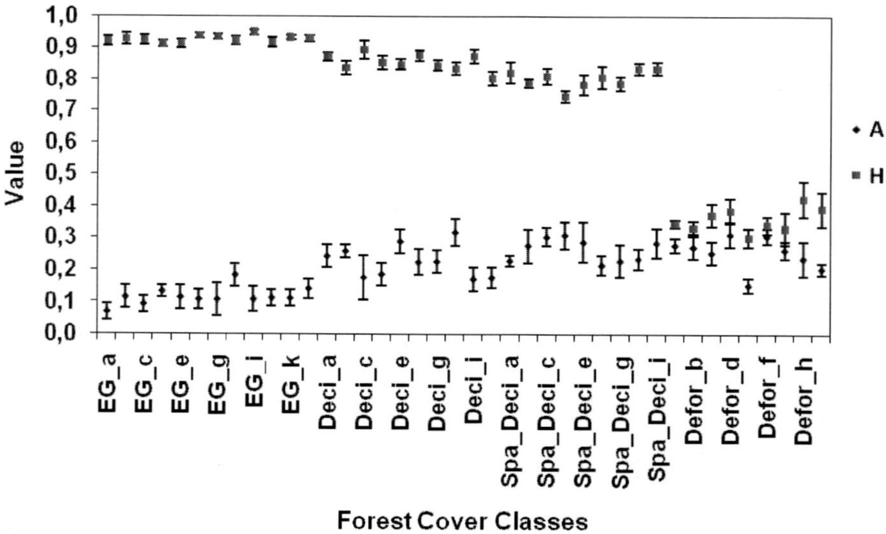

Figure 4.9. Variations of the Entropy (H) and Anisotropy (A) with forest cover classes.

Figure 4.10 shows the RVI image of the study area. The mean value of RVI is higher for evergreen forest (0.91) compared to those shown by deciduous (0.67) and sparsely deciduous (0.54) (Figure 4.11). High value of

RVI in evergreen forest is because of the multi-story tree structure with complex canopy which causes high σ^o in HV polarization. The RVI shows a good relation with woody vegetation (Ling et al., 2009). The value of RVI is lower for sparsely deciduous forests as compared to that of deciduous due to the low volume scattering. However, RVI is very low for deforested area (0.17) because of lower HV backscattering.

The estimated pedestal height (PH) is shown in figure 4.12. The mean value of PH is observed to be higher for evergreen, (0.41) medium for deciduous (0.26) and sparsely deciduous (0.20) forests, and lower (0.04) for the deforested area (Figure 4.13). PH gives a good contrast between different types of forest-cover. The High value of PH indicates more un-polarized scattering components in the received signals and the presence of multiple scattering mechanisms. PH is a good parameter for differentiating the types of the forests and the deforested area.

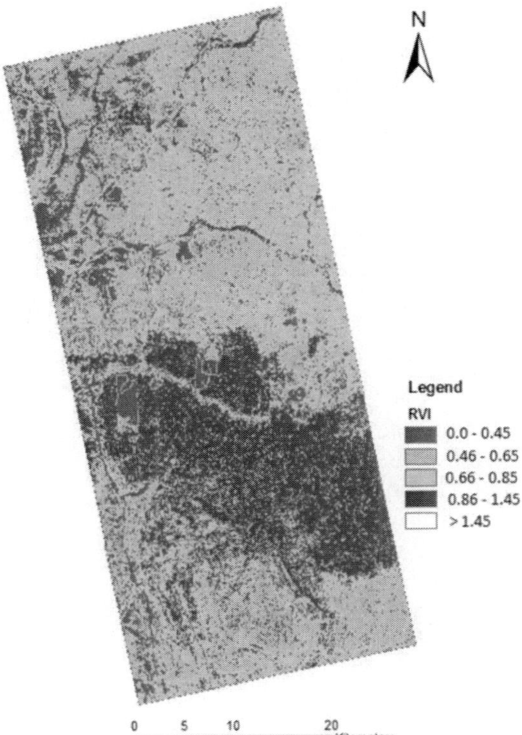

Figure 4.10. Radar Vegetation Index (RVI)image of the study area.

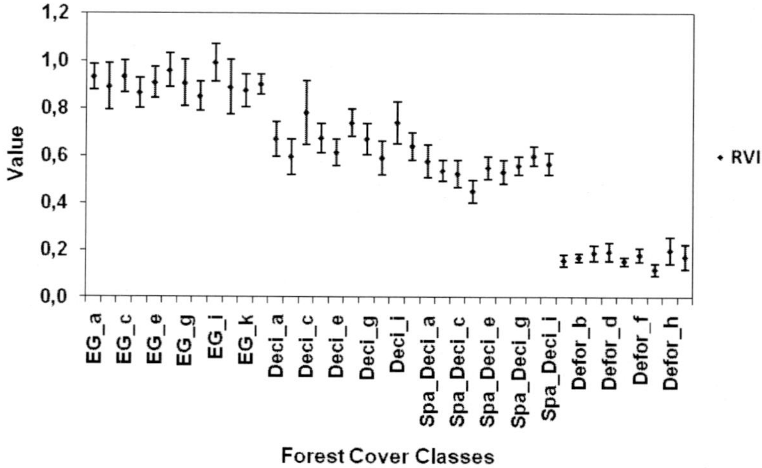

Figure 4.11. Variations of the RVI with forest cover classes.

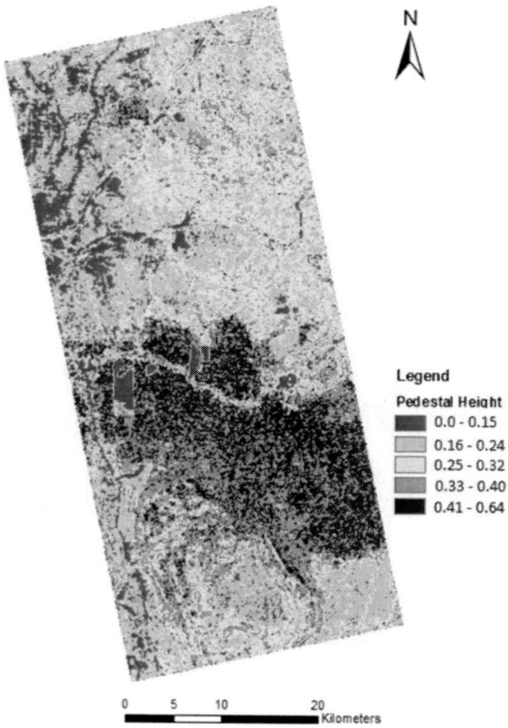

Figure 4.12. Pedestal Height (PH) image of the study area.

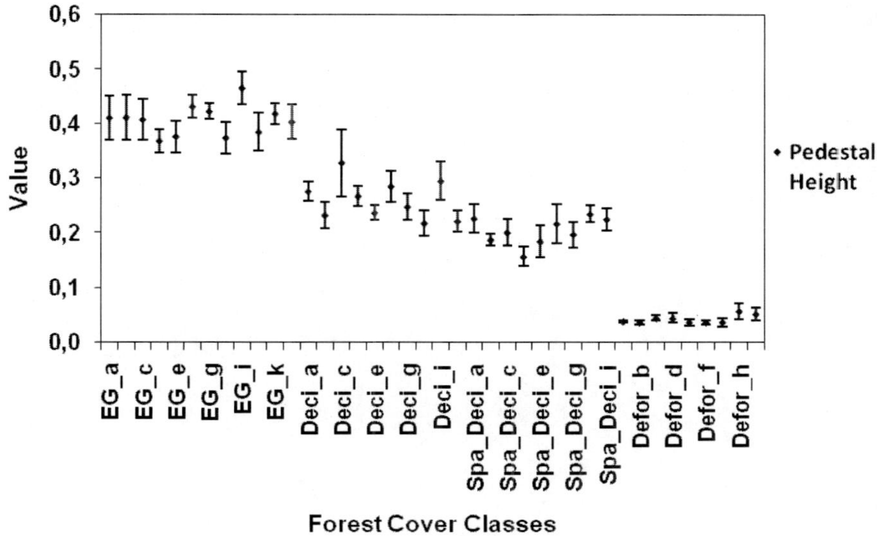

Figure 4.13. Variations of the pedestal height (PH) with forest cover classes.

Figure 4.14 shows the R:G:B colour composite of the Freeman-Durden decomposition. It gives realistic and detailed information about the types of scattering mechanisms for various scatterers, and provides the percentage of even, odd and double bounce scattering mechanisms occurring in the study area. The evergreen forest shows a higher value of volume scattering (0.43) compared to the deciduous (0.42) and the sparsely deciduous (0.31). In contrast, the evergreen forests shows a low value of surface (0.03) and double bounce scattering (0.01), whereas deciduous and sparsely deciduous forest shows a high value(Figure 4.15, Table 4.2). The deforested area shows a very low value of the volume (0.04) and double bounce (0.01) scattering and high for surface scattering (0.25). High volume scattering in evergreen forest is because of the multi-story tree structure with complex canopy (Guerra et al., 2008).

Figure 4.16 shows the variation of value of volume, surface, double bounce and helical scattering from various forests cover classes. The evergreen forest shows higher volume scattering than those of deciduous and sparsely deciduous forests. The evergreen forest shows low values for surface and double bounce scattering, whereas deciduous and sparsely deciduous forests show a higher surface and double bounce scattering. The deforested area shows very low value of volume and double bounce scattering however, the value of surface scattering is high for deforested area. Yamaguchi four component based decomposition is useful for forests types' classification and deforestation monitoring.

Hence this chapter shows that Freeman-Durden three component and Yamaguchi four component based polarimetric decomposition is useful for identification of forests and deforested area. However, use of Yamaguchi four component based decomposition is more significant because of wide use in forests cover classification (Chapter 3) and deforestation study.

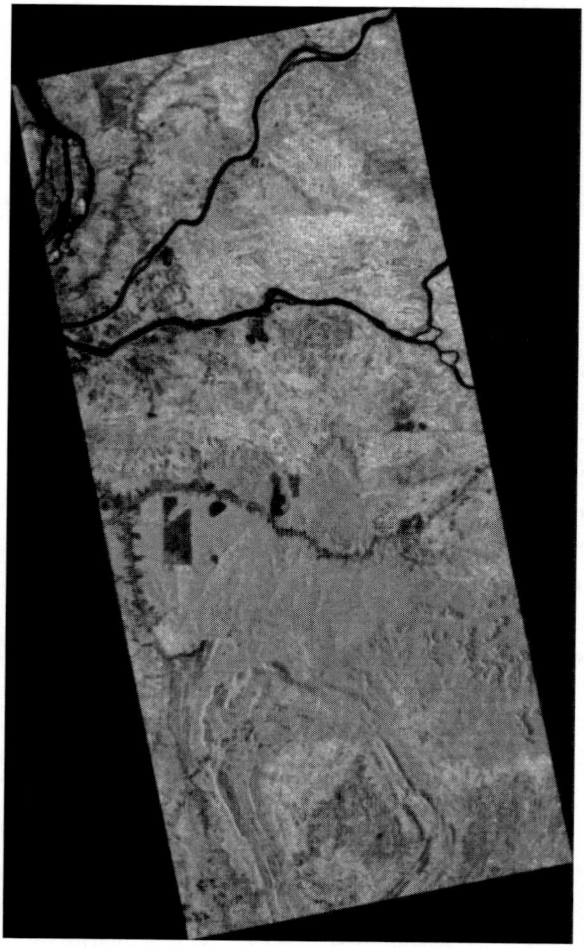

Figure 4.14. Freeman-Durden (R:Double bounce scattering, G:Volume scattering, B:Surface scattering) image of PALSAR data.

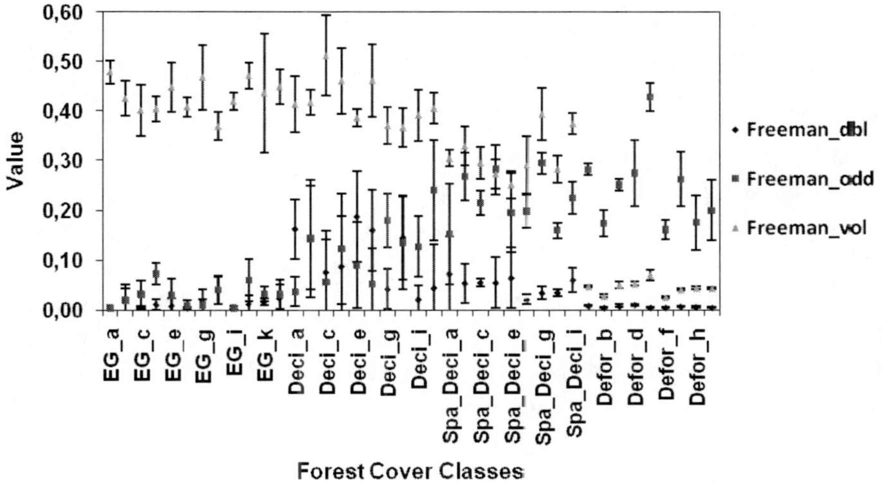

Figure 4.15. Variations of the Freeman scattering with forest cover classes.

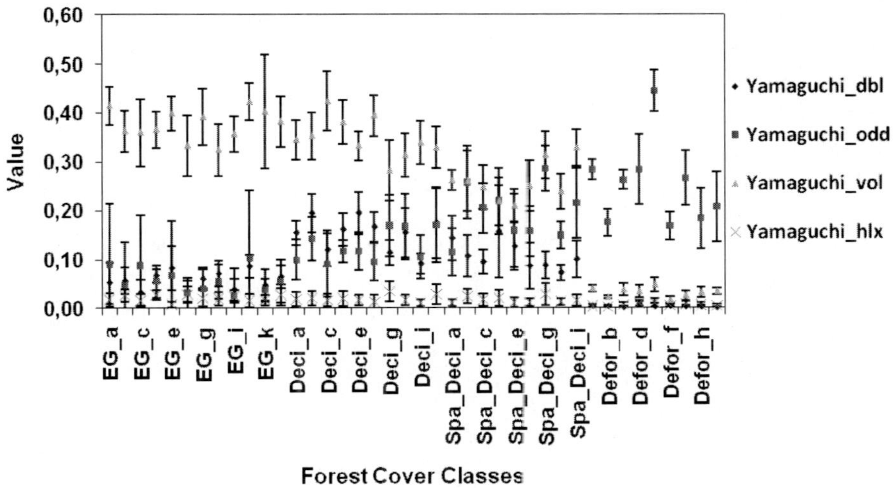

Figure 4.16. Variations of the Yamaguchi scattering with forest cover classes.

CONCLUSION

Characterization of the forests and deforestation in Cambodia have been made using full polarimetric PALSAR by observing the backscattering coefficients (σ^o), H, α, A, PH, RVI, Freeman-Durden three-component and Yamaguchi four component based decomposition parameters. These properties depend on the interaction of PALSAR signal with forest types. The mean σ^o HV observed for evergreen, deciduous, sparsely deciduous and deforestation were -12.96, -12.99, -14.29, and -22.25 (dB), respectively. The deciduous forests have shown the value of σ^o HV close to evergreen forest may be because of saturation of PALSAR signal at 100 tons/ha as noticed by Watanabe et al., 2006. The value of mean σ^o HV is very low for deforested area due to the absence of volume scattering. The observed entropy and RVI were highest for evergreen forest and lowest for the deforested area, whereas the pattern was reverse for HH/HV polarization ratio. Freeman-Durden three component and Yamaguchi four component based decomposition shows high volume scattering in forested area whereas surface scattering contribute in deforested area. This chapter shows that σ^o HV, cross-polarization ratio (HH/HV), H, RVI, PH and Yamaguchi four components based decomposition gives better results for characterization of forest-types and deforested area as compared to other polarimetric parameters. The advantage of this chapter is to identify deforested area during rainy season when optical sensors cannot provide cloud free images. Use of all these polarimetric parameters can provide better deforestation monitoring as compared to σ^o HV only.

Chapter 5

DEFORESTED AREA HEIGHT ESTIMATION: DEMS DATA

5.1. INTRODUCTION

The estimation of carbon (C) loss due to deforestation activities required for REDD+ carbon accounting. However, development of cost-effective methodology is still a challenge because of the uncertainties in estimation of forest volume. If REDD+ signed countries are reducing their carbon stocks by the deforestation activities, they need to monitor the loss of carbon stocks (biomass loss) for carbon credits. For that, information about the forest volume is required. Currently, most of the present research is focused on the development of a reliable and cost-effective forest volume estimation technique using the satellite data.

Previous studies have shown that forest canopy height, and above ground biomass (AGB) are strongly related (Zhao et al., 2009; Ni-Weister et al., 2010) and forest height is needed for the estimation of volume of the deforested area. Most of the optical and Synthetic Aperture Radar (SAR) satellites provide only the spatial and temporal distribution of forest canopy but not vertical distribution (Sun et al., 2008). Although the new Polarimetric Interferometric Synthetic Aperture Radar (PolInSAR) and Light Detection and Ranging (LiDAR) techniques can measure the forest height, but each of them has some limitation. PolInSAR has temporal and volume decorrelation due to the time difference between two acquisitions cause changes in the geometric and dielectric properties of objects (Papathanassiou and Cloude, 2003; Eriksson et al., 2008) and thereby affects coherence. In most of the previous studies airborne based PolInSAR data was used to minimize the temporal decorrelation and found a good correlation with PolInSAR derived forest height and biomass (Mette et al., 2003; Hajnsek et al., 2009). However, PALSAR based PolInSAR techniques have been facing problem of temporal and volume decorrelation because of the long repeat cycle (46 days) and airborne based PolInSAR cannot provide global information and expensive. Airborne LiDAR system is the best available technique for the estimation of height and 3-D structure of the forest, but it is costly and of low spatial coverage (Zhao et al., 2009, Asner et al., 2010). Recent high resolution SAR (e. g. TerraSAR-X, COSMO-SkyMed and TanDEM-X) can overcome temporal and volume decorrelation, but it is also very expensive. Because of the above discussed limitations of present remote sensing techniques, alternative techniques to estimate height of cut over forests are highly required.

In this chapter, an attempt to estimate the height of cut over forest in Cambodia using the GLAS, SRTM, GDEM and PRISM-DSM data in a cost effective way has been made. The height of cut over forest is useful parameter to know the 3-D information about forests that will be useful to calculate carbon loss due to deforestation activities.

5.2. Study Area

In Cambodia, logging activities are mainly occurring in mature forest, which is locking up more carbon stocks compared to other forests (Sasaki and Yoshimoto, 2010). About 44% (4.7 million ha) of total forests of Cambodia are managed by concession companies and play a vital role in economic development of Cambodia (Kim Phat et al., 2001). The testy site of this study lies in the Stung Treng and Kratie province of Cambodia (Figure 5.1). It covers approximately 1340 km^2 of area.

5.2.1. Land Use Land Cover Map of the Study Area

Land-use/land-cover map of the area was generated using the ASTER data acquired in 2005, implying the maximum likelihood classification techniques, supported with ground truth observation (Figure 5.2). The test site was classified into 12 classes, namely, the deciduous forest, evergreen forest, agricultural land, water and other classes. Most of the study area is covered by evergreen forest (43%), followed by deciduous (19%) and agriculture land (19%) (Table 5.1).

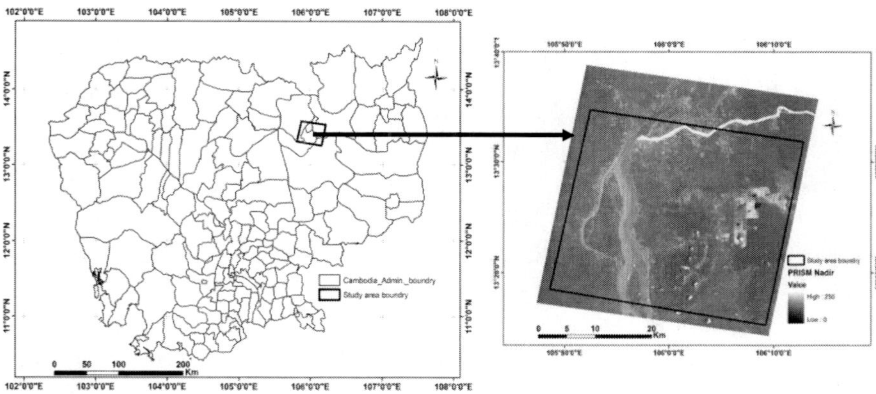

Figure 5.1. Location of the study area with ALOS/PRISM Nadir view data.

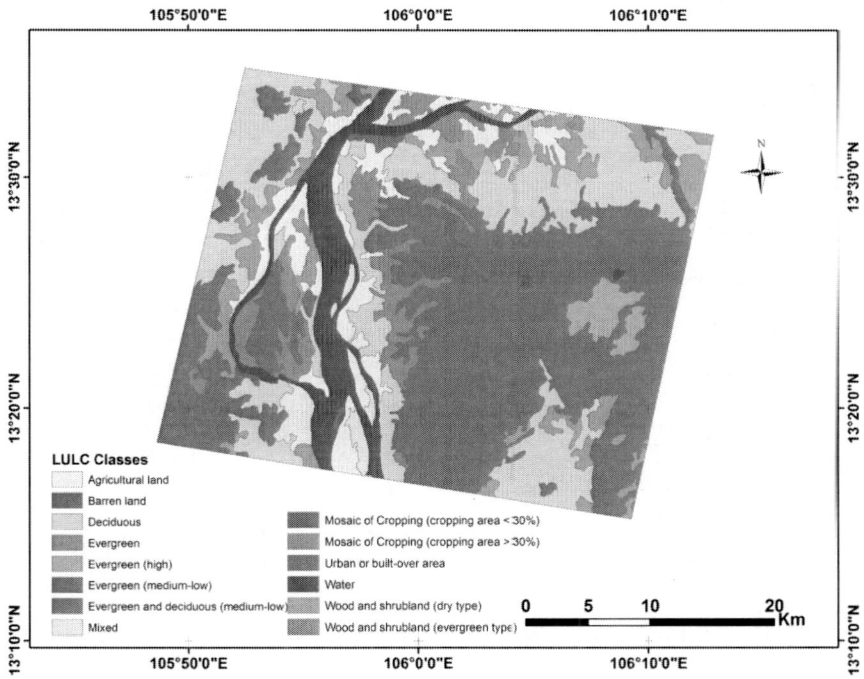

Figure 5.2. Land use Land cover map of the area based on supervised classification of ASTER 2005 data.

5.3. METHODOLOGY

5.3.1. Satellite Data

5.3.1.1. Glas

Measurement by the GLAS is ranged from GLA 1 – GLA 15. However, in this research, we used GLA14 product acquired on May 19 (L3A, release 28) and October 04 (L3E, release 31) in 2004, and February 23, 2006 (L3J, release 31) and February 18, 2008. GLA14 is a product of global land surface altimetry, provides the surface elevation, including the geo-location of the footprints. According to Sun et al., (2008), the waveform of the GLAS with a size of 65 m footprint is similar and highly correlated with 10 m footprint size of the Laser Vegetation Imaging Sensor (LVIS). Claudia et al., (2006) suggested that global elevation of the GLAS can be used to evaluate the accuracy of SRTM-DEM. The GLAS data has been widely used for mapping the forest canopy, can measure the height at an accuracy of about 10 cm in flat and non-vegetated surfaces (Fricker et al., 2005; Martin et al., 2005; Chen, 2010, Lefsky et al., 2005 and Simard et al., 2008), validating the estimation of SRTM (Bhang et al., 2007; Sun et al., 2008) and other applications. Yamanokuchi et al., (2007) also used GLAS data for correction of DEM generated from InSAR. Therefore, the GLAS data is used as a reference to correct other DEM data for this research, since previous studies suggested that it is more accurate - spaceborne LiDAR.

Table 5.1. Percentage of various land use/land cover classes

No.	LULC Classes	Area (Km^2)	Percentage
1	Agricultural land	259.4	19.36
2	Barren land	0.5	0.04
3	Deciduous	259.4	19.36
4	Evergreen	50.7	3.79
5	Evergreen (high)	21.0	1.57
6	Evergreen (medium-low)	452.0	33.74
7	Evergreen and deciduous (medium-low)	52.5	3.92
8	Mixed	18.2	1.36
9	Mosaic of Cropping (cropping area > 30%)	26.3	1.97
10	Water	95.9	7.16
11	Wood and shrub land (dry type)	101.6	7.58
12	Wood and shrub land (evergreen type)	2.1	0.16
	Total	1339.7	100.00

5.3.1.2. SRTM-DEM

SRTM-DEM data with a resolution of 90 m (three-arc second) is used to estimate the surface elevation in the study area for the year 2000. SRTM DEM data was generated from the fixed baseline interferometric C-band SAR in February, 2000 (Sarabandi and Lin, 2000). SRTM-DEM also represents the digital surface model (DSM) because most of the study area was covered by the forest during the year 2000 (figure 5.5a) and C-band SAR cannot penetrate the forest canopy.

5.3.1.3. ASTER GDEM Data

ASTER GDEM data with 30 m resolution (1-arc second) is also additionally used for estimation of elevation. GDEM was generated in 2009 from stereo-pair images acquired with nadir and backward images of ASTER over the same area.

5.3.1.4. PRISM-DSM Data

ALOS/PRISM level 1B2 data with 2.5 m resolution with a nadir and backward view was acquired on November, 11th, 2009. The RPCs (Rational Polynomial Coefficients) information of both scenes was also used in this study for the generation of DSM.

Table 5.2. Forest's inventory data for evergreen and deciduous forest

No.	Forest Types	Mean Height (m)	Mean DBH (cm)	Tree density (no./ha)	Plot Size (ha)	Biomass (t/ha)
1	Evergreen_1	17	19.7	544	0.18	170
2	Evergreen_2	20.3	21.96	555	0.18	254
3	Evergreen_3	18.3	25.39	338	0.18	480
4	Deciduous_1	10.5	20.6	222	0.9	79.7
5	Deciduous_2	7.5	21.2	311	0.9	99.7
6	Deciduous_3	10.3	25.5	166	0.9	99.3
7	Deciduous_4	13.5	19.7	577	0.9	149.3
8	Deciduous_5	12	21.6	238	0.9	141

5.3.2. Field Data

The forest inventory and ground truth data were collected from the study area in December 2009 and January 2011. Three plots in evergreen forest and five plots in deciduous forest were selected based on Landsat satellite spectral signature (visual interpretation of Landsat false color composite) for forest inventory data collection. The 30x60m and 30x30m plot size forest inventory data comprising the Diameter at Breast Height (DBH), tree height, and species were collected. Details are summarized in Table 5.2. These field data were used to validate the results.

5.3.3. Data Processing

GLAS data is processed using the standard methodology based on the IDL code provided by the National Snow and Ice Data Center (NSIDC). After processing, the GLAS data was projected into the UTM, WGS 84 projection system. The 90 m SRTM-DEM and 30 m ASTER-GDEM data were re-sampled into 10 m resolution using nearest neighborhood method. In this chapter, DSM data with 10 m resolution was generated from the PRISM stereo images for deforested area from nadir and backward view with RPCs information. According to Saunier, (2007), PRISM-DSM generated from the nadir and backward scenes have least position errors compared to other combinations of look angle. The stereo photogrametry based PRISM-DSM has some effects of clouds and the quality of DSM is degraded because of clouds. The elevation data derived from PRISM-DSM has less accuracy because of clouds and RPCs information. Therefore, the correction is needed after generation of all DEMs. The offset value was calculated for each DEM based on GLAS elevation data. As have been mentioned, GLAS data has been taken as reference data. Agriculture area has been selected as a suitable site for the comparison of all DEMs because agriculture area is flat area and very small change occurred in elevation as compared to other land cover types. Figure 5.3 shows comparative values of elevations of various DEMs for the river, evergreen forest, deciduous forest and agriculture area. Table 5.3 illustrates the statistical average, standard deviation (SD) and offset values of various DEMs used for different classes of land covers. There is a difference in elevation in each DEM, because the data and processing steps are different for each type of DEM. The calculated offset value is useful for comparative study of the four DEMs. After calculation of offset, offset corrected DEMs has been generated and results show that the DEMs after correction are better than before. The deforested area height map was generated by subtracting the PRISM-DSM offset corrected data with those of SRTM-DEM offset corrected data. The detailed methodology flow chart is given in Figure 5.4. Landsat 2010 data has been used to digitize the deforested area based on visual interpretation.

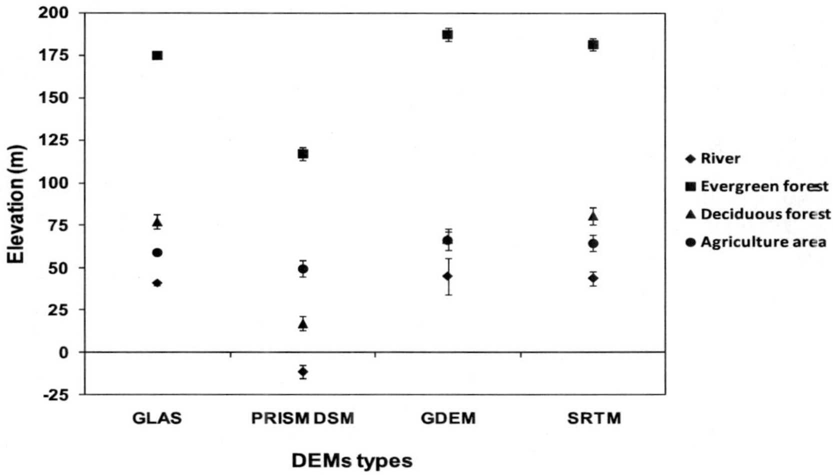

Figure 5.3. Elevation data of the four DEMs types for different land cover classes.

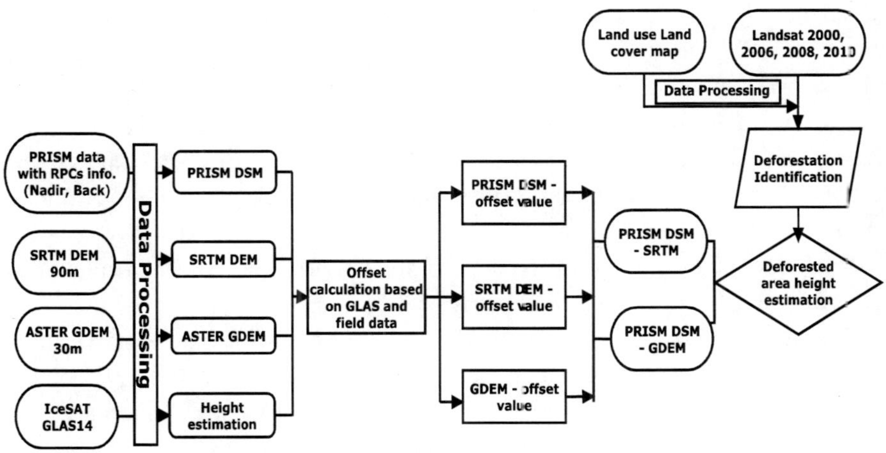

Figure 5.4. Flow chart of the methodology adopted.

Table 5.3. Calculation of offset value of different DEMs

	GLAS (m)		PRISM DSM (m)			ASTER GDEM (m)			SRTM DEM (m)		
	Mean	St. Dev.	Mean	St. Dev.	Offset	Mean	St. Dev.	Offset	Mean	St. Dev.	Offset
River	40.8	1.1	-11.2	2.9	52.0	45.1	12.8	-4.3	43.8	2.3	-3.0
Evergreen Forest	174.8	2.2	116.7	2.8	58.0	187.3	3.1	-12.4	181.5	1.7	-6.7
Deciduous Forest	76.9	4.2	16.9	5.2	60	66.8	4.2	10.1	80.6	3.9	-3.7
Agriculture area	58.7	1.3	49.6	4.7	9.1	66	5.3	-7.3	64.6	5.0	-5.9

5.4. Results and Discussion

Figure 5.5 (a-d) shows the Landsat R:G:B false color composite (FCC) images for the year December 2000, June 2006, March 2008 and February 2010, respectively. These Landsat images clearly show the deforestation trend. Landsat ETM+ image of December 24, 2000 shows the homogeneous intact evergreen forest (Figure 5.5a), but due to logging activity this intact forest started degradation, evidently can be seen in the Landsat TM image in 08th June 2006 (Figure 5.5b). This deforested area was assigned to a private concession company, which started deforestation for rubber and paulownia plantation (Field based information). Those deforested area can be easily identify in Figure 5.5c by visual interpretation. Figure 5.5d shows the deforested area has been increased with the time because of high deforestation activity. Based on visual interpretation vectorization of Landsat images of year 2010 (Figure 5.5d), the deforested area of about 122.5 km^2 has been calculated in the test site.

Figures 5.6, 5.7 and 5.8 show the SRTM-DEM, GDEM and PRISM-DSM maps with a resolution of 10 m of the area overlayed with GLAS shots, indicating the range of elevations from 31-240 m, out of which the elevation range of 151-180 m, 151-180 m, and 91-120 m respectively, are showing the elevation in the deforested area. Low elevated areas are present along the river side. The elevation maps obtained from the SRTM-DEM, GDEM and PRISM-DSM are not showing same elevation as these were generated using different data sources and methodology. Table 5.4 illustrates the limitation and resolution properties of various DEMs. To minimize the effects of various processing techniques, GLAS elevation data has been used as a reference and calculated offsets value for each DEM. The offset corrected DEMs showed better elevation than previous DEMs. These offset corrected DEMs were further used to estimate the height of the deforested area. The offset corrected PRISM-DSM is used as a base, because PRISM-DSM data is most recent, generated in 2009.

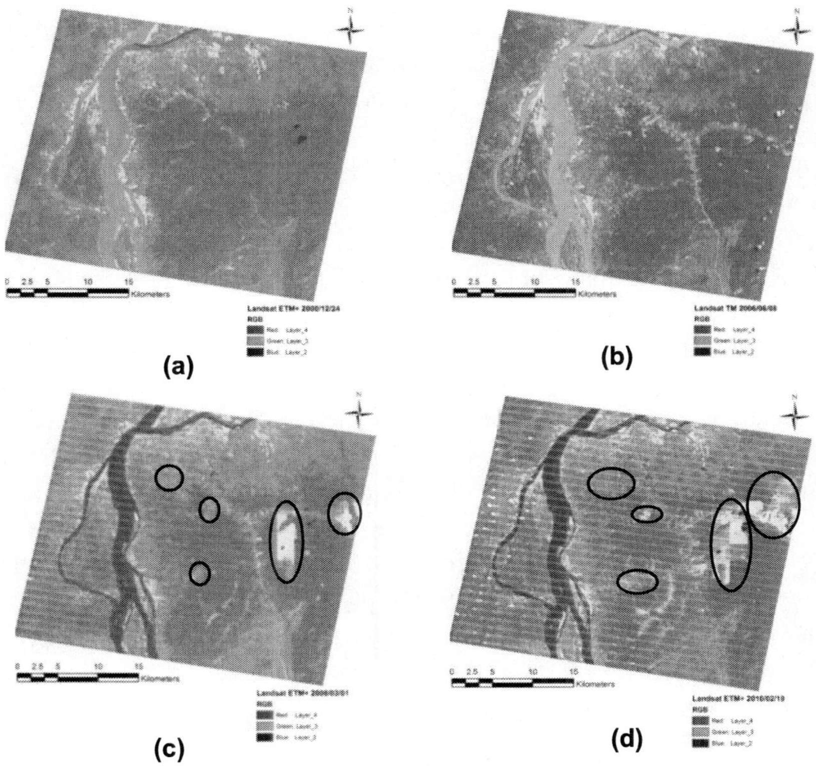

Figure 5.5. Landsat R:G:B (band-4:band-3:band-2) false color composite images of year (a) 2000 (b) 2006 (c) 2008 and (d) 2010. Black circle is showing deforested area.

The offset corrected GDEM and SRTM are differentiated with that of offset corrected PRISM-DSM. Figure 5.9 shows the elevation difference between offset corrected PRISM-DSM and offset corrected GDEM, while Figure 5.10 shows an elevation difference between offset corrected PRISM-DSM and offset corrected SRTM-DEM. The difference values from both the maps for deforested land were from -29 to -10 m (Figure 5.9) and -29 to -20 m (Figure 5.10), respectively. If we compare the difference values, from offset corrected PRISM-DSM and offset corrected SRTM-DEM (Figure 5.10) is more reliable and promising than those from the offset corrected PRISM-DSM and offset corrected GDEM (Figure 5.9). If we compare both the maps (Figure 5.9 and Figure 5.10) the deforested lands are more clearly seen in figure 5.10 than in figure 5.9 (visual comparison). Hence, PRISM-DSM and SRTM-DEM derived results are better than that of PRISM-DSM and GDEM. Similarly, other land cover classes (deciduous forest, agriculture land, etc.) are also showing the difference from -9 to 5 m for most of the area. If we compare the deforested area height (-29 to -20m) with those of surrounding forests (-14 to -5m), we found that there is a difference of about 15 m of height. This difference of elevation between deforested and forested area represents the height of the cut over forest, consistent with the average forest height of 17, 20.3 and 18.3 m from forest inventory data of three evergreen plots (Table 5.2). Figure 5.11 shows the deforested area overlaid on Landsat data of 2010. Deforested area was digitized based on visual interpretation of deforested area using Landsat data. This figure shows that most of the deforested area falls into those of PRISM and SRTM (-29 to -20 m). In some places PRISM-DSM and SRTM derived deforested area has discrepancies because of the clouds cover and other reasons, for example, the difference in the observation period of Landsat data (February 2010) and PRISM-DSM (November 2009). In some instances, the PRISM-DSM and SRTM cannot identify the degraded forests. Above discussed study is useful for the estimation of volume of the deforested area since the calculation of height of cut over forests using this methodology is more reliable. The validation result shows ±5 m of uncertainty in forest height when compared with forest inventory data. Forest height measured during field survey at the precision of ±0.5 to 2 m using hypsometer and remaining uncertainty is because of the PRISM data processing using RPCs information. Takaku et al., (2004) has also indicated that absolute accuracy of PRISM DSM is less than 5 m.

Table 5.4. Comparison between DEM data

	SRTM DEM	**ASTER GDEM**	**PRISM DSM**	**ICESat GLAS**	**LiDAR**
Limitations	Steep terrain, low resolution	Clouds	Clouds, RPCs information	Clouds, slope, Large foot-print size	Clouds, expensive
Resolution	90 m	30 m	2.5-10 m	65 m	50 cm or less
Vertical accuracy	10-20 (m)	7 -14 (m)	5 (m)	20 cm	10 cm

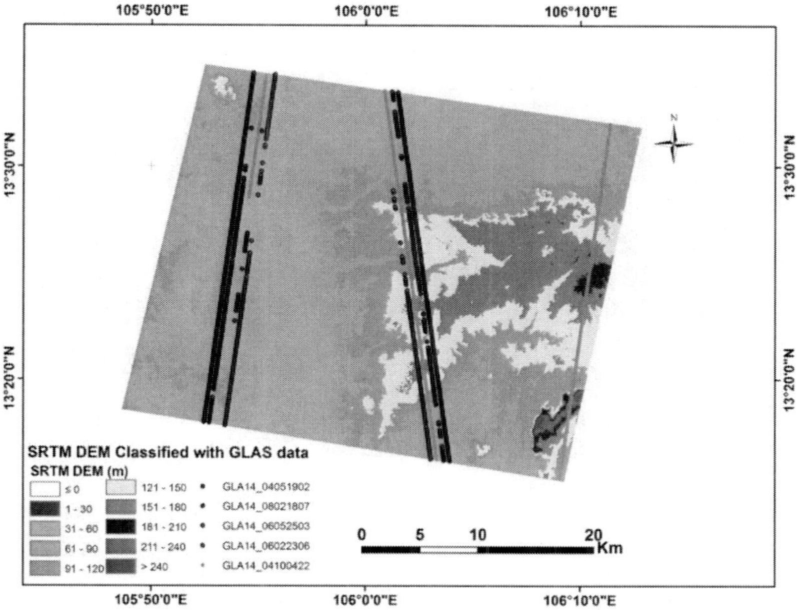

Figure 5.6. STRM DEM data of the area with ICESat GLAS shots.

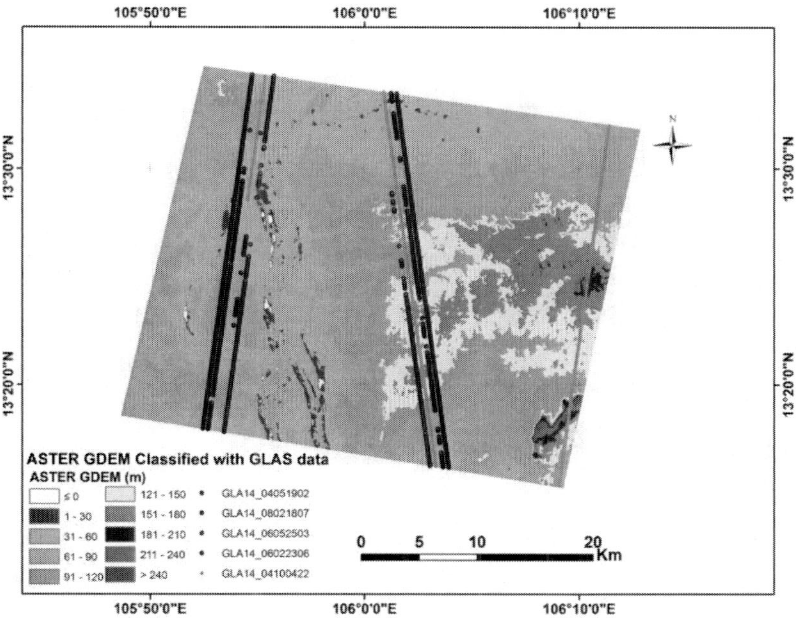

Figure 5.7. ASTER GDEM data of the area with ICESat GLAS shots.

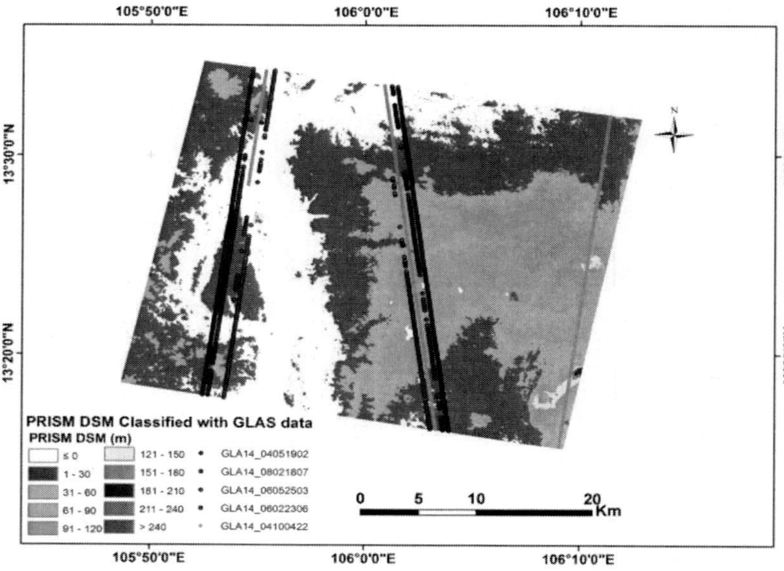

Figure 5.8. PRISM DSM data of the area with ICESat GLAS shots.

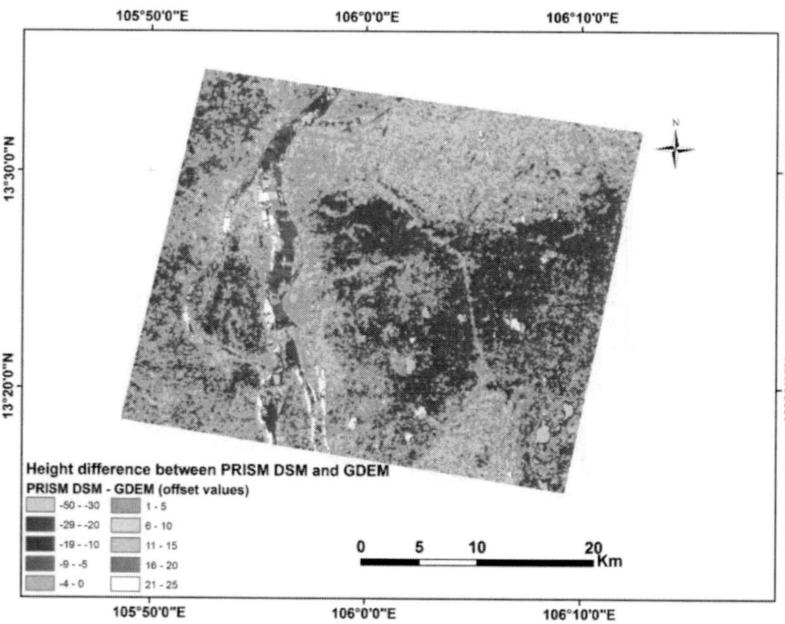

Figure 5.9. Difference between offset corrected PRISM DSM data and offset corrected GDEM data.

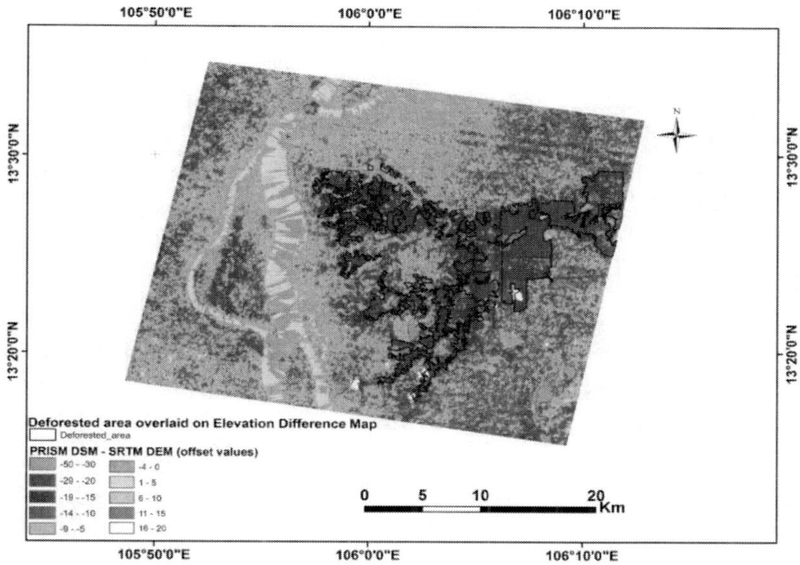

Figure 5.10. Difference between offset corrected PRISM DSM data and offset corrected SRTM DEM data (deforested area overlaid on elevation difference map).

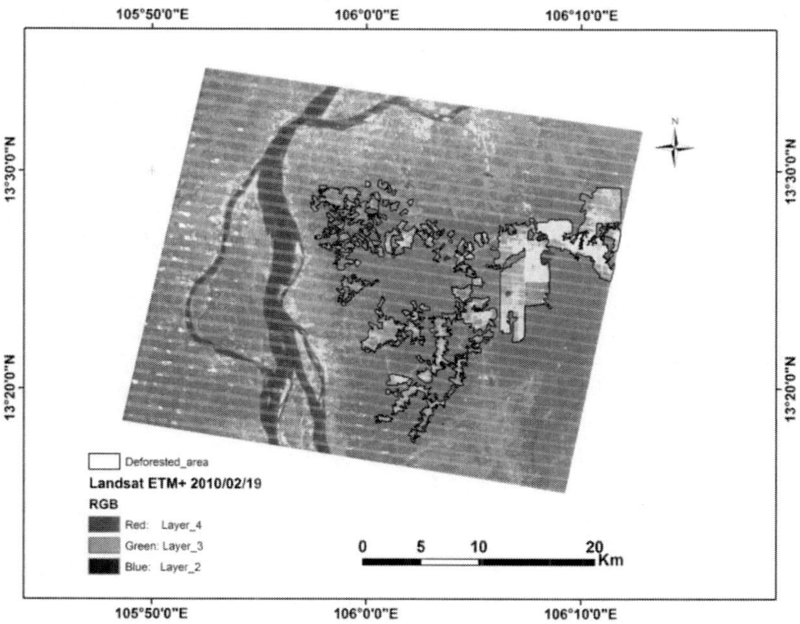

Figure 5.11. Landsat 2010 data overlaid with deforested area.

CONCLUSION

Accurate estimation of forest volume is needed for the calculation of forest carbon stocks and absorption of carbon from the atmosphere. To estimate forest volume, height is a key parameter and needs to be estimate cost-effectively for the UN-REDD+. Height of cut over forests has been estimated based on the GLAS, SRTM-DEM, ASTER-GDEM and PRISM-DSM. The estimated height for deforested area using the offset corrected PRISM-DSM minus offset corrected SRTM-DEM and offset corrected PRISM-DSM minus offset corrected GDEM were varied from -29 to -20 m and -29 to -10 m, respectively. But the result from the offset corrected PRISM-DSM minus offset corrected SRTM-DEM is more reliable than those from the offset corrected PRISM-DSM and offset corrected GDEM because of lack of exact acquisition date information of ASTER images to generate ASTER-GDEM. Cut over forests height map shows consistency with the forest inventory. Hence, this study is useful for the estimation of height of cut over forests in a cost effective way, with some limitations such as: (a) Ineffectiveness of PRISM-DSM data during cloudy days, (b) lack of date of acquisition of ASTER-GDEM generation (c) limited GLAS shots in study area (d) detection of height of deforested area with a size less than 3x3 pixels of SRTM data, which are beyond this methodology.

Chapter 6

PALSAR 50M MOSAIC DATA BASED NATIONAL LEVEL BIOMASS ESTIMATION

6.1. INTRODUCTION

Forests biomass information is useful for REDD+ carbon accounting and trading carbon credits (Wang et al., 2011). Forest carbon pools consist of trunks, branches, leaves, litter, dead wood, roots and soil carbon. However, most of the studies have focused on above ground biomass (AGB) because this is relatively large pool and other carbon pools can be calculated based on simple equations (Ravindranath and Ostwald, 2008). Table 6.1 shows the distribution of carbon among different pools in forests and other wooded lands (FAO, 2006).

There are various methodologies for the biomass estimation but no methodology presents a clear view on how to report carbon pools and their fluxes and what would be accuracy and uncertainty of biomass monitoring. Therefore biomass mapping has become an urgent need to assess and produce data on forest carbon stocks and change in stocks on a national level (Maniatis and Mollicone, 2010; Wijaya, 2010). A recent biomass map Sacchi et al., (2011) shows uncertainties of about 30-50% in Indo-China countries. Table 6.2 shows the previous studies related to forest biomass estimation.

The most accurate way of calculating biomass is destructive sampling and forest inventory data using allometric equations. However, these traditional techniques are often time consuming, labor intensive, and difficult to implement, especially in remote areas and they cannot provide the spatial distribution of biomass in large areas. Moreover, the method cannot give historical information about forest if there were no existing forest inventory data (Lu et al., 2005; Sivanpillai et al., 2006). Therefore satellite remote sensing supplemented with low forest inventory data can provide cheap and fast estimation as well as historical information about forest biomass. Most of the remote sensing techniques are based on optical and synthetic aperture radar (SAR) systems. The disadvantages of optical sensors are not related to the plants structural parameters, acquisition of cloud free images in tropical countries, low saturation level of spectral bands and various indices (Gibbs et al., 2007). Therefore, dependency on SAR sensors has been increased because SAR can provide data without limitation of clouds and solar illumination. Penetration capability of SAR allows to provide information about plants structural parameters hence have ability to measure biomass (Lu, 2006).

The successful launch of the Advanced Land Observing Sattelite's (ALOS) PALSAR in 2006 has increased the potential to use radar to measure biomass, as this is the first long-wavelength (L-band, 23-cm wavelength) SAR satellite sensor to have the capability of collecting single, dual, full and Scan-SAR mode with cross-polarized (HV, horizontal-transmit, vertical receive) and co-polarized (HH, horizontal-transmit, horizontal receive; VV, vertical-transmit, vertical receive) data. The HV polarization is useful because it interacts with trees and produces a strong response (Mitchard 2011). Various studies have analyzed the retrieval of AGB using radar data in tropical regions (Mitchard et al., 2009; Gama et al., 2010; Djomo et al., 2011; Englhart et al., 2011). Longer wavelengths SAR have proven to be more useful because of an increasing backscatter range with changing biomass (Dobson et al., 1992; Luckman et al., 1997; Castro et al., 2003; Lu, 2006). These biomass estimations are valid up to a certain threshold where saturation occurs, (Lucas et al., 2007; Mitchard et al., 2009). In general, saturation level of SAR depends on the frequency of SAR systems as well as the forests structure. The sensitivity of SAR decreases with the increase of biomass in the dense forests (Imhoff, 1995; Kasischke et al., 1997). Currently there is no comprehensive study that uses PALSAR 50m mosaic data to generate national level biomass map. Therefore this study was carried out to estimate national level biomass based on the bottom-up approach to support REDD+ policies in Cambodia.

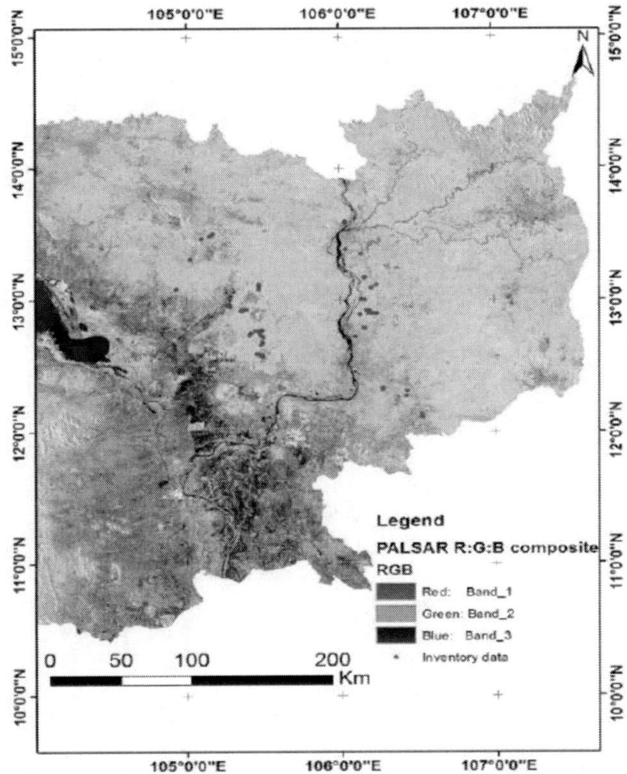

(a) Evergreen forest

(b) Mixed forest

(c) Deciduous forest

Figure 6.1. ALOS/PALSAR 50m mosaic 2009, (Red: HH, Green: HV, Blue: HH/HV) data and locations of the inventory data in different forest types (a) evergreen (b) mixed and (C) deciduous forests of Cambodia.

6.2. Study Area

Cambodia has one of Asia's highest rates of forest loss. It has lost 29% of its primary forests during 2000 to 2005. The deforestation rate is highest in Cambodia among Indo-China countries (FAO, 2010). Logging activities, population growth, urbanization, and agricultural expansion has been the primary reason for Cambodia's forest loss (UNEP, 2009; Gaughan et al., 2009). Cambodia has signed REDD+ policies in 2009 and study of forest biomass is necessary for REDD+ policies implementation. Figure 6.1 shows R:G:B colour composite of PALSAR 50m FBD data.

Table 6.1. Distribution (%) of carbon among different pools in forests and other wooded lands (FAO 2006) (Figures in parentheses are percentages of the row totals)

No.	Region/ sub-region	Carbon in				
		Living biomass	Deadwood	Litter	Soil	Total
1	East and South Africa	63.5	7.5	2.1	-	73.1
2	North Africa	26.0	3.3	2.1	33.5	64.9
3	West and Central Africa	155.0	9.8	2.1	56.0	222.9
4	Total Africa	95.8 (59.5)	7.6 (4.6)	2.1 (1.6)	55.3 (34.3)	160.8 (100.0)
5	East Asia	37.0	5.0	-	-	41.9
6	South and South-east Asia	77.0	9.0	2.7	68.4	157.1
7	West and Central Asia	39.0	3.6	11.4	41.0	95.8
8	Total Asia	57.0	6.9	2.9	66.1	132.9
9	Total Europe	43.9 (24.8)	14.0 (7.9)	6.1 (3.4)	112.9 (63.9)	176.9 (100.0)
10	Caribbean	99.7	8.8	2.2	70.5	181.2
11	Central America	119.4	14.4	2.1	43.3	179.2
12	North America	57.8	8.8	15.4	35.8	117.8
13	Total North and Central America	60.1	9.0	14.8	36.6	120.6
14	Total Oceania	55.0	7.4	9.5	101.2	173.1
15	Total South America	110.0	9.2	4.2	71.1	194.6
16	World Average	71.5 (44.4)	9.7 (6.1)	6.3 (3.9)	73.5 (45.6)	161.0 (100.0)

Table 6.2. Previous studies related to forest biomass estimation

No.	Authors	Study Area	Methodology	Data used
1	Sader et al., 1989	Luquillo Mountains, Puerto Rico	Normalized Difference Vegetation Index (NDVI)	Landsat MSS and TM, Simulator, airborne multispectral scanner
2	Beaudoin et al., 1994	Landes Forest, France	Adapted theoretical model	SAR P-band
3	Rauste et al., 1994	Freiburg, south-east Germany; Ruotsinkyla, Finland	Linear regression analysis	AIRSAR C,L,P band
4	Brown et al., 1995	Rondonia State, Southwestern Brazilian Amazon	Allometric equation based on destructive sampling approach	Field measurement data
5	Imhoff, 1995	Hawaii Volcanoes National Park	multi polarization (HH, HV, VV) radar backscatter (σ)and polynomial regression model	NASA/JPL Airsar data with C, L, and P band
6	Harrell et al., 1997	South – eastern USA	Multiple regression analysis	SIR-C
7	Luckman et al., 1998	Tapajos, Para state and Manaus, Amazonas state, Brazil	Forest backscatter model	JERS-1 SAR L band
8	Steininger, 2000	Bolivia and Brazil	TM band 3,4,5 validated with allometric equations	Landsat TM
9	Austin et al., 2003	New South Wales, Australia	Linear regression analysis	JERS-1 SAR L band
10	Santos et al., 2003	Tapajos River region, Para state, Brazil	Regression models (logarithmic and polynomial function)	AeS-1 SAR P- band
11	Foody et al., 2003	Manaus (Brazil), Danum Valey (Malaysia) and Khun Khong (Thailand)	vegetation indices, complex band ratios complemented with multi-linear regression and neural networks method	Landsat TM

Table 6.2. (Continued)

No.	Authors	Study Area	Methodology	Data used
12	Lu, 2005	Eastern Brazilian Amazon: Altamira, Pedras, and Bragantina	LandsatTM bands, vegetation indices, band ratios, image transform (e.g., principal component analysis, Tasseled cap)	Landsat TM
13	Kuplich et al., 2005	Manaus and Tapajos forests, Brazil	Radar backscatter (σ) and GLCM texture based allometric equations	JERS-1 SAR image with L band
14	Watanabe et al., 2006	Temperate Coniferous forests	Multi-linear regression	PALSAR
15	Sales et al., 2007	Rondonia State, Southwestern Brazil	Stem volume – AGB equation and kriging method	Field data (RADAMBRASIL database)
16	Hajnsek et al., 2009	Mawas and Sungai Wain, Kalimantan, Indonesia	RVoG model and inversion of dual-polarization	Airborne multi-band (C, L, P, X band) and multi-polarization (PolInSAR)
17	Lucas et al., 2010	Queensland, Australia	Regression modelling	PALSAR
18	Sun et al., 2011	Boreal forests of Howland, Maine (US)	Multi-linear regression analysis	LVIS and PALSAR
19	Sandberg et al., 2011	Hemiboreal forest, Sweden	Regression modelling	L-band and P-band SAR data
20	Saatchi et al., 2011	Tropical forests	Regression modelling	GLAS, MODIS, SRTM and QSCAT
21	Englhart et al., 2011	Tropical forest on Central Kalimantan, Indonesia	Regression modelling	TerraSAR-X and PALSAR
22	Mitchard et al., 2011	Central Africa (central Cameroon)	Regression modelling	PALSAR

6.3. METHODOLOGY

6.3.1. Field Data

Forest inventory data was collected in November 2010 and January 2011 with plot size of 30x60m. Square plot design was used to facilitate pixel sampling based on satellite data to reduce position error. Systematic random sampling design was applied for the purpose of field data collection based on forest types with relatively homogeneous ecological conditions (i.e., topography, slope, distance from water source, soil types). Forest inventory parameters (Diameter at breast height (DBH), tree height, species, tree density and forest types) were collected from seventy nine plots. The tree diameter was measured at 1.3m height above the ground using DBH tape with 1cm accuracy. The tree height was measured with the vertex hypsometer for all open trees with good visibility of the top and was estimated when it was not possible to see the top of the tree. Most of the sampling plots were selected in the plane area to minimize topographic effects of SAR data. The sampling plots were located using GPS (Garmin 62CSx). The AGB in kg for each tree was calculated using allometric equation as derived by Kiyono et al., (2010).

Table 6.3. Allometric Equations

No.	Author	Allometric equation
1	Kiyono et al., 2010	Leaf biomass (kg) = $173*(BA^{0.938})$ Branch biomass (kg) = $0.217*(BA^{1.26})*(D^{1.48})$ Stem biomass (kg) = $2.69*(BA^{1.29})*(D^{1.35})$ BA is basal area and D is stem density
2	Kenzo et al., 2009	Leaf biomass (kg) = $0.0442*(DBH^{1.67})$ Branch biomass (kg) = $0.0124*(DBH^{2.48})$ Stem biomass (kg) = $0.0822*(DBH^{2.48})$
3	Brown et al., 1997	Biomass (kg) = $(42.69-12.8*DBH+1.242*(DBH^2))$

This study used Kiyono et al., (2010) allometric equations for biomass calculation based on forest inventory data because Anglesen (2008) has mentioned that country specific allometric equation is better than global allometric equation. We also compared the Kiyono et al., (2010) allometric equation based biomass with the Brown (1997) and Kenzo et al., (2009) allometric equations (Table 6.3) based biomass then we found that the Brown (1997) allometric equation based biomass estimation showed over estimation and Kenzo et al., (2009) showed under estimation. The biomass value obtained from each tree using allometric equation were summed and normalized by the area of the plots to produce the AGB in Mg/ha. In this biomass estimation, we have considered the tree ≥ 10cm DBH, which are likely to comprise of most of the woody mass of the plots.

6.3.2. Satellite Data

Land use/land cover map based on ASTER 2005 data, SRTM-DEM data, Landsat ETM+ 2009, 2010 data were used for selection of sampling sites. PALSAR FBD 50m mosaic data was downloaded from Japan Aerospace Exploration Agency (JAXA). The processing of PALSAR data was started with the terrain corrections using Shimada et al., 2010 methodology to minimize the topographic effects of PALSAR in mountainous area. The terrain correction methodology was not very effective because of highly rugged topography in the area. The digital number (DN) was converted to the normalized radar cross section (NRSC or σ^o). The backscattering coefficient was calculated using the following equation (Shimada 2009).

$$\sigma^o = 10 \times \log_{10} (DN^2) - 83 \quad (1)$$

The PALSAR data was co-registered with Landsat ETM+ orthorectified data. We have not considered the climatic conditions of PALSAR 50m mosaic data because this information was not available.

6.3.3. Statistical Analysis

Multi-linear regression (MLR) analysis using the stepwise forward method was conducted relating the σ^o of PALSAR to corresponding field calculated biomass. 3x3 pixels window size analysis has been used for MLR model development and this MLR model has been applied to the PALSAR 50m mosaic data to estimate biomass of whole Cambodia. Finally validation was used to evaluate the accuracy of the model by comparing PALSAR estimated AGB to the field derived AGB. The detailed methodology is shown in the flow chart (Figure 6.2).

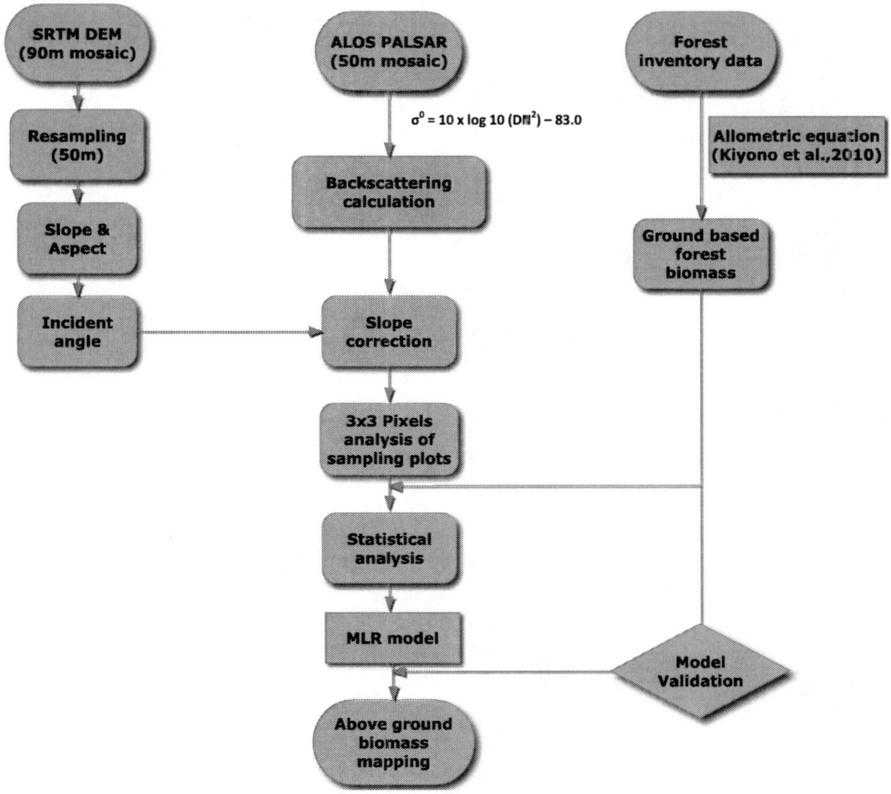

Figure 6.2. Flow chart of the methodology.

6.4. Results and Discussion

Statistical analysis has been made to relate the forest inventory data. Figure 6.3a, b and c show the relationship of biomass with the stem density, average DBH and basal area respectively. Figure 6.3a does not show good correlation (R^2 = 0.2) between biomass and tree density because the tree density depends on the tree species, site conditions, forest types etc. Figure 6.3b does not show good correlation (R^2 = 0.003) between biomass and average DBH because the average DBH is the average of all DBH in particular plot and depends on tree density and forest types. Figure 6.3c shows strong correlation (R^2 = 0. 9) between biomass and basal area because basal area is a function of tree density and DBH. A total seventy nine plots have been analyzed. Fifty one plots were used for the MLR model development and 23 plots were used for model validation. Five plots were excluded from the analysis because the location of the plots was near to the road as well as due to some degradation activity.

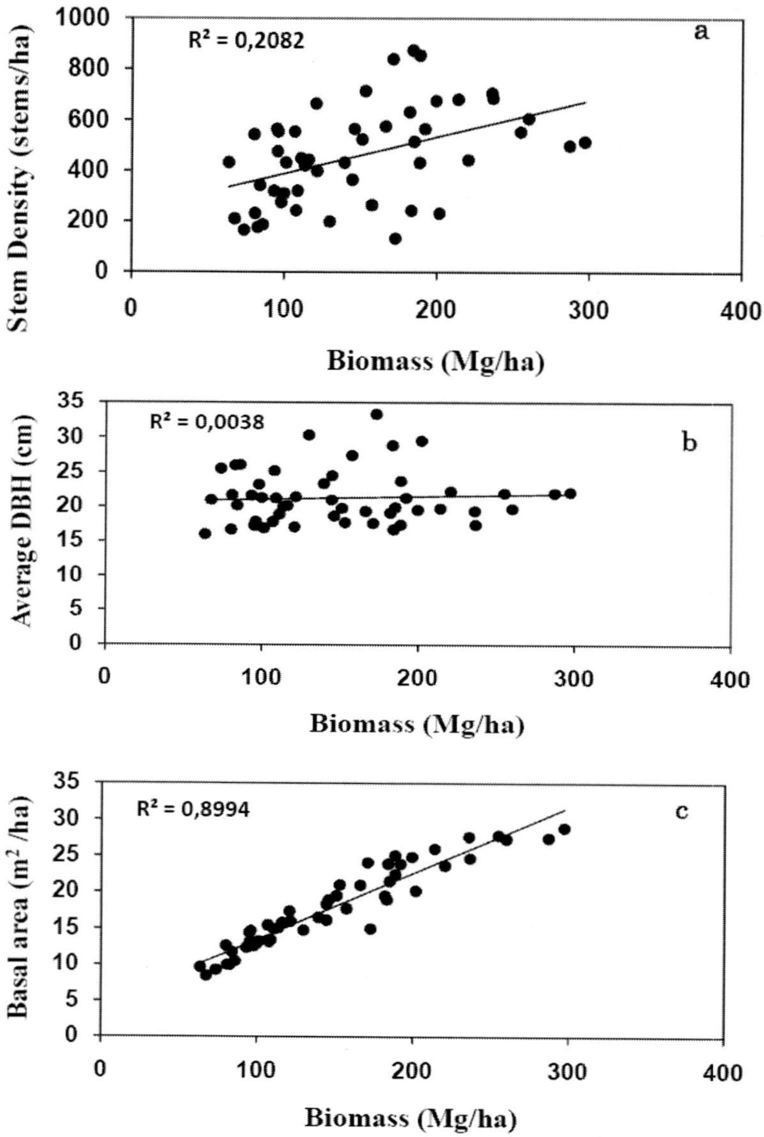

Figure 6.3. Biomass value for the field plots against (a) stem density (b) average DBH and (c) basal area.

σ^o HH, HV and HH/HV is plotted against basal area, stem density and biomass respectively. Figure 6.4 a and b represents the relationship between PALSAR σ^o (HH, HV) and (HH/HV) with basal area. Field measured basal area shows a significant relationship with the σ^o HV ($R^2 = 0.67$) as compared to σ^o HH ($R^2 = 0.05$). Figure 6.4 c and d represents the relationship between PALSAR σ^o (HH, HV) and (HH/HV) with stem density. Field measured stem density shows the poor relationship with the σ^o HV ($R^2 = 0.32$) and σ^o HH ($R^2 = 0.06$) respectively. This is mainly because tree density depends on the forest types, tree species and site conditions. Figure 6.4d also shows poor relationship between σ^o HH/HV with tree density ($R^2 = 0.3$). Figure 6.4e shows the relationship between PALSAR σ^o (HH) and (HV) with biomass. Field measured biomass shows a significant relationship with the σ^o HV ($R^2 = 0.67$) as compared to σ^o HH ($R^2 = 0.05$). High σ^o HH in low biomass region has been noticed because of the high surface scattering from the plots covered by dry leaves and grasses, which increases the surface roughness. The reason why σ^o HV polarization produces better correlation than σ^o HH is because of the volume scattering in forest areas enhances the cross-polarization returns as increase in biomass. σ^o HV is less influenced by soil and vegetation moisture than σ^o HH (Collins et al., 2009) as well as VanZyl (1993) has noticed that HV is less influences by topography. Other studies also reveal that the σ^o HV is more sensitive to the forest biomass compared to σ^o HH (Le Toan et al. 1992; Harrell et al. 1995; Mitchard et al., 2009). We have observed different backscattering properties from the same biomass region (Figure 6.4e at biomass 100-150 Mg/ha) because of difference in canopy and their distribution. Evergreen forest having multi-story tree structure shows high backscattering as compared to deciduous forests of same biomass class. A loss in sensitivity of PALSAR signal has been appeared to occur at approximately 150-200 Mg/ha biomass (Figure 6.4e). Figure 6.4f shows the strong relationship between PALSAR σ^o HH/HV with biomass ($R^2 = 0.56$). Therefore, polarization ratio is useful parameter for biomass estimation. These results have a higher saturation point and less noise as compared to previous studies suing PALSAR HV data (Watanabe et al., 2006; Mitchard et al., 2009). Similar saturation point using PALSAR HV data has been noticed by Mitchard et al., (2011) in Savanna forest of central Africa. The factors to achieve this range of saturation could be good quality, well goecoded forest inventory data in relatively flat areas which produced more accurate results than other previous studies.

MLR Model for biomass estimation was developed using σ^0 HV and σ^0 HH/HV because HV and HH/HV shows strong correlation with biomass. σ^0 HH data were not used for regression modelling, since the dependence on biomass was weak. The resulting regression model is given in equation 2. σ^0 HV is dominated by volume scattering from woody elements of trees, so that HV is strongly related to AGB (Le Toan et al., 2011). For the HH polarization, ground conditions can affect the biomass backscattering relationship, because HH backscatter comes mainly from trunk-ground scattering (Ranson et al., 1994).

$$Y \text{ (Biomass (Mg/ha))} = 752.95 + (44.4 \times \sigma^0 \text{HV}) - (8.73 \times \sigma^0 \text{HH/HV}) \tag{2}$$

The MLR model has been applied to the PALSAR 50m mosaic data to generate national level biomass map. Figure 6.5a shows the biomass map of the Cambodia. The biomass values were classified into 8 classes. Deforested area shows the zero biomass value. Figure 6.5b shows the land use land cover (LULC) map of the same biomass region. If we compared the biomass map (Figure 6.5a) with the LULC map (Figure 6.5b) then the high biomass region (>200 Mg/ha) was mostly falls into the evergreen high and medium low class of the LULC map. However, in mountainous area (northern part) the biomass map shows variation because of the topographic effects. Low biomass region (150-200 Mg/ha) was mostly found in mixed forest type and lowest biomass region (50-150 Mg/ha) was mostly found in deciduous forests. The results from this study are preliminary, but it shows the potential of PALSAR 50m mosaic data that is freely available.

Figure 6.6 shows the validation results of PALSAR derived biomass. The accuracy of PALSAR predicted AGB decreases as the biomass increases because of saturation of PALSAR signal. It shows a significant coefficient of correlation $R^2 = 0.61$. The overall root mean square error (RMSE) for these data is 63 Mg/ha; however this decreases to 19 Mg/ha if only values below 100 Mg/ha are considered and to 21 Mg/ha using values up to 200 Mg/ha. The high variation in errors are present in the high biomass region i.e., >200 Mg/ha. We have predicted two types of uncertainties a) calculating biomass from field data using allometric equation because we have not used species specific allometry as well as small plot size and exclusion of small trees (< 10 cm) DBH and b) saturation of PALSAR signal at high biomass region as well as topographic effects.

Figure 6.7 illustrates the national level biomass distribution, which indicates high heterogeneity in biomass class. The high biomass region mainly occurred in Ratnakiri, Mondolkiri, Kampong Thom and Koh Kong province of Cambodia. Figure 6.8 summarizes the distribution of biomass in various types of forests of Cambodia. Evergreen forests have highest biomass in the >200 Mg/ha class, whereas deciduous forests have in the 100-150 Mg/ha class. The largest biomass stock is in evergreen forests followed by deciduous forests. The high biomass of evergreen dense forests may be associated with good environmental factors such as relatively better water, soil and temperature conditions and less intensity of human activities. Some sites in evergreen forests of Kampong Thom province shows deforestation may be because of strong human activities in plane area with high biomass forest.

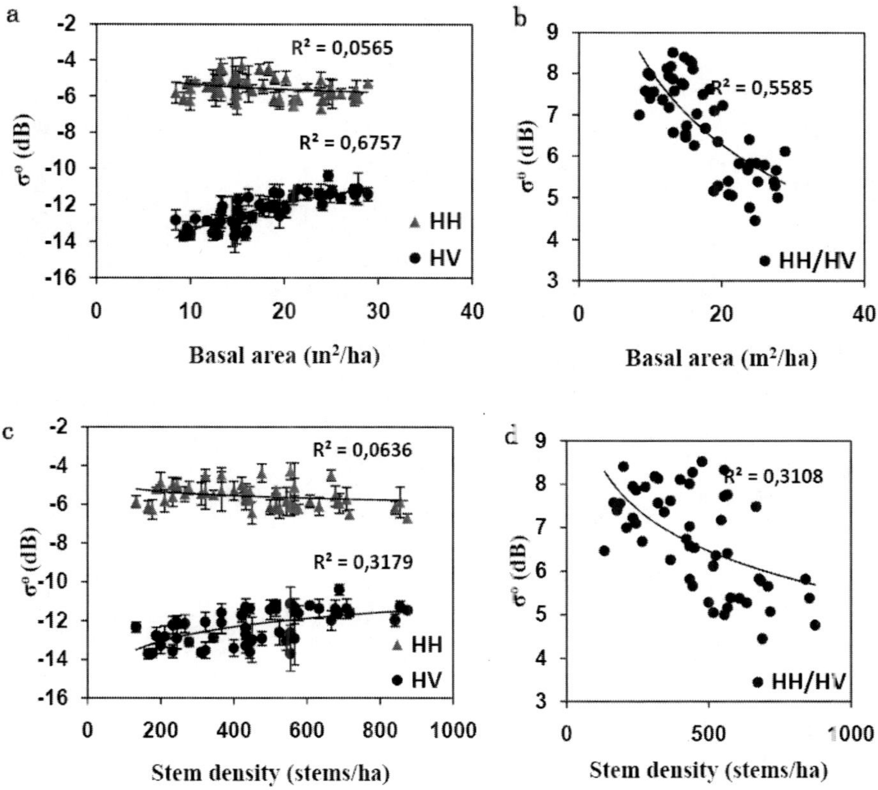

Figure 6.4. (Continued).

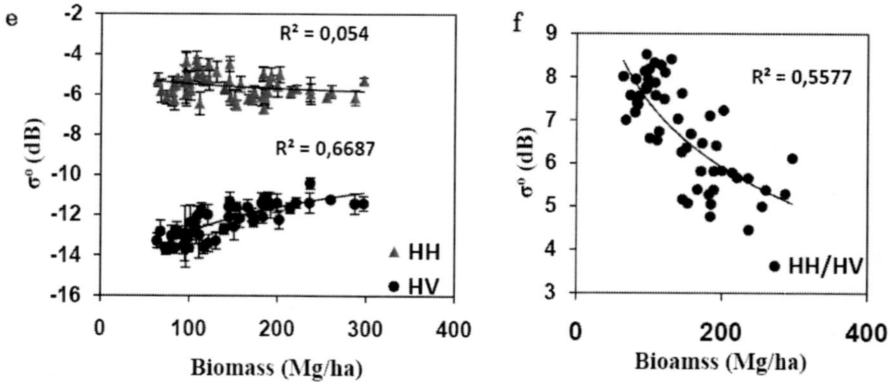

Figure 6.4. PALSAR 2009 σ^0 HH, HV and HH/HV plotted against biomass (a, b), stem density (c, d) and basal area (e, f).

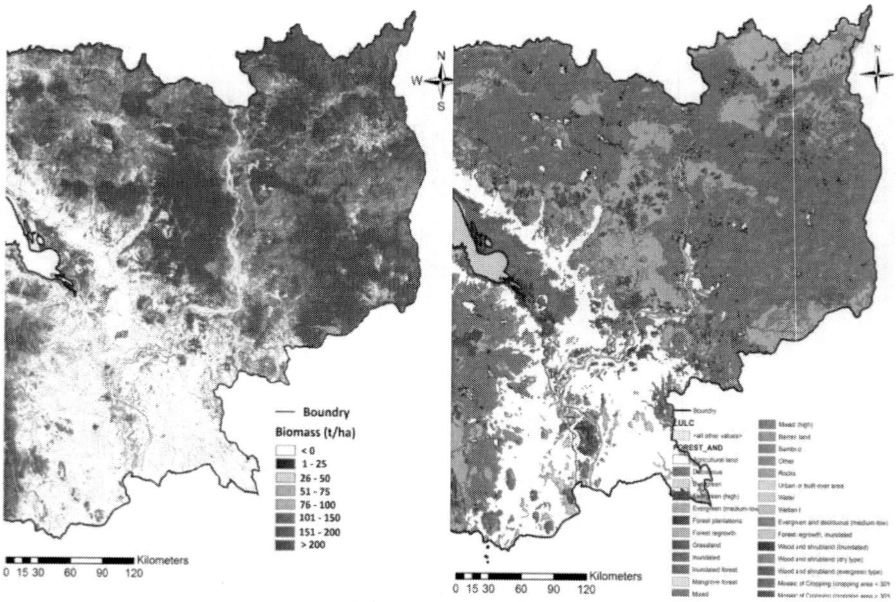

(a) (b)

Figure 6.5. (a) PALSAR derived biomass map (b) LULC map of the area.

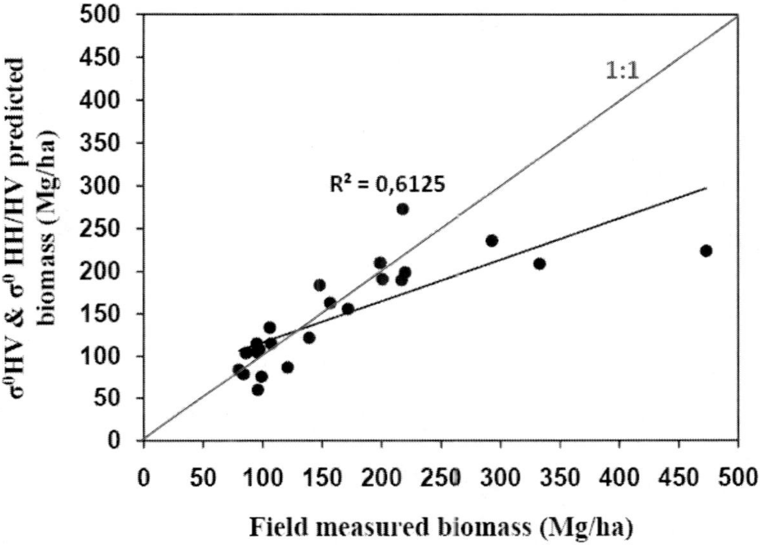

Figure 6.6. Relationship between PALSAR predicted biomass plotted against field measured biomass [RMSE = 63 Mg/ha].

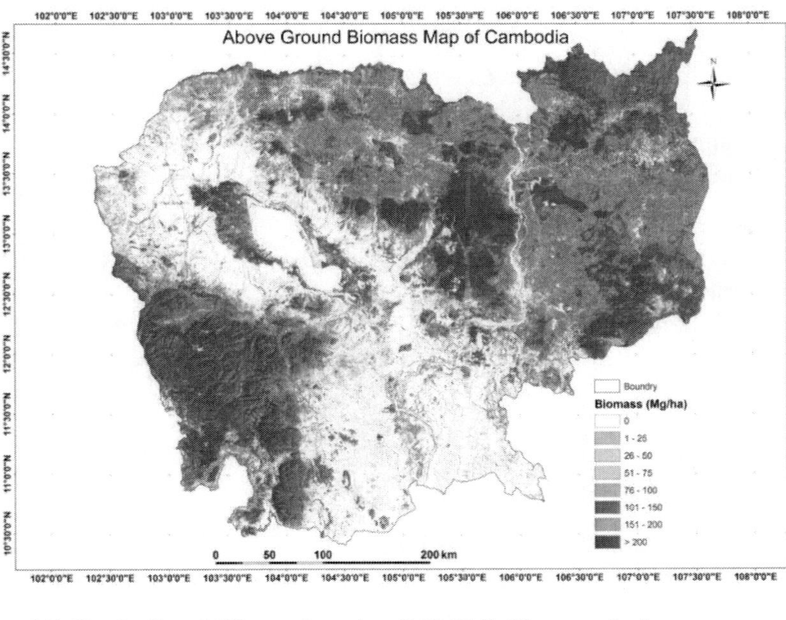

Figure 6.7. Cambodian AGB map based on PALSAR 50m mosaic data.

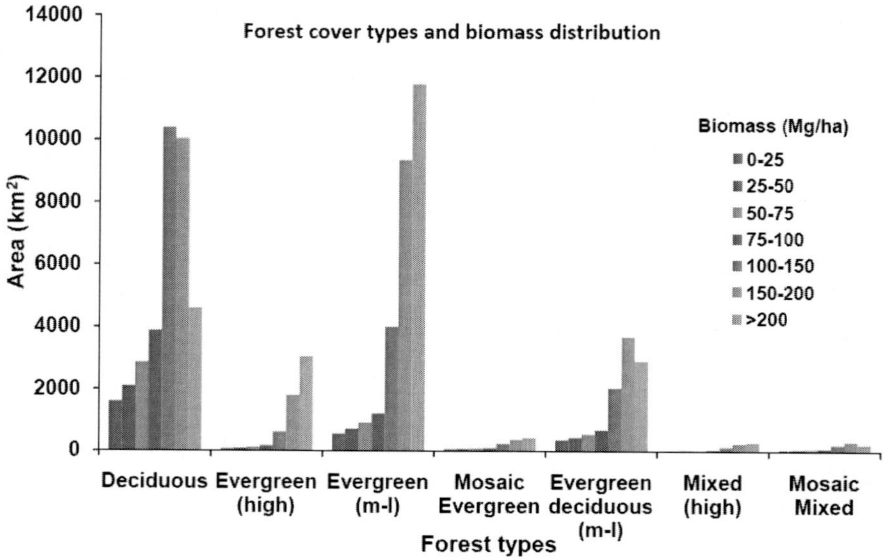

Figure 6.8. Biomass distributions with forest cover types of Cambodia.

Figure 6.9 shows the pie chart which depicts the percentage area of forest and their biomass class. The majority of Cambodia's forests fell in a range of 100-200 Mg/ha (~ 52%) and only 20% of total forest area were < 100 Mg/ha. About 28% of the total Cambodia's forests having value >200 Mg/ha, which is highly significant in the climate change mitigation point of view. Conservation of these high biomass forests should be on priority to increase the carbon stock as well as for biodiversity conservation point of view.

We also compared our results (Table 6.4) with the Sato, (2011) estimated biomass value (Table 6.5) that was generated based on 100 permanent sampling plots (PSPs) in evergreen and deciduous forests of Cambodia. We estimated total AGB of evergreen forests is about 347 ± 140 Tg-C, which is very similar to values presented by Sato, (2011). The value of carbon stock in deciduous forests is about 238 ± 71 Tg-C, which is also very close to the Sato, (2011) estimated carbon stock value. Therefore we could say that the PALSAR 50m mosaic generated biomass is reasonable and can be use for further studies. However, more accurate biomass map is really needed for further accurate climate change modelling.

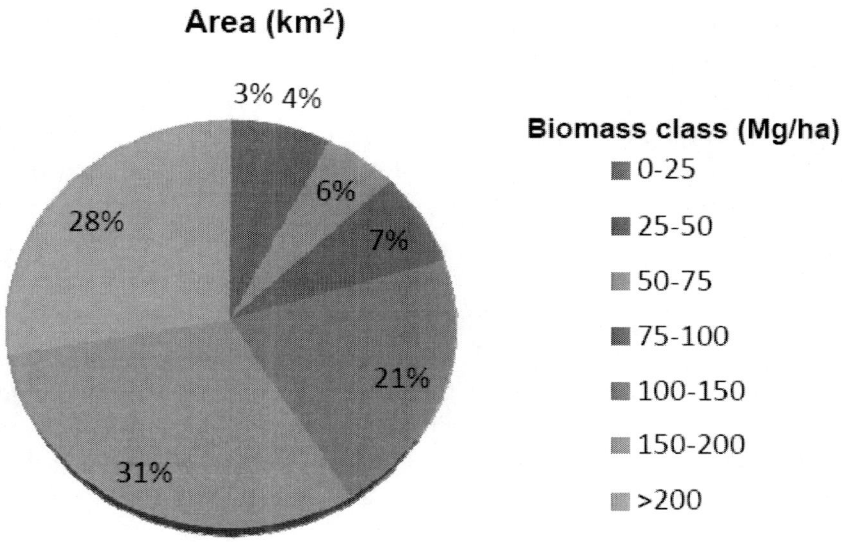

Figure 6.9. Biomass distributions with percentage of forest cover area.

Table 6.4. Cambodian forest carbon stock based on ALOS/PALSAR 50m

Forest type	Forest area (km^2)	Total biomass (Mg/ha)	Total Carbon stock (Tg-C) with 30% uncertainties
Evergreen forest	36140.27	694839050	347.42 ± 140.2
Deciduous forest (It does not include mosaic deciduous forest)	35729.55	477427500	238.71 ± 71.6
Mixed forest	12588.67	205317175	102.66 ± 30.8
Total	84458.49	1377583725	688.79 ± 206.7

Table 6.5. Forest carbon stock in Cambodia (Sato, 2011)

Forest type	Forest area (km^2)	Average carbon stock (2000-2001) (Mg-C/ha)	Total Carbon stock (Tg-C)
Evergreen forest	36,689	127 ± 79.4	467.2 ± 291.5
Deciduous forest	46,921	33.7± 23.6	158.2 ± 110.8
Total	83,610		625.4 ± 219.5

This national level biomass map (Figure 6.7) will assist Cambodian forestry administration, land managers, policy makers and civil society to become more informed about the likely result of their policies and program in reducing national GHG emissions from land use change. This biomass map can be used as an additional tool for forest conservation and forest management strategies of Cambodian government.

CONCLUSION

Biomass information is useful for calculation of amount of carbon loss due to deforestation activity. In this chapter, a method for estimating national level biomass map using PALSAR 50m mosaic data has been developed and evaluated. Correlation analysis was used to assess the relationship of AGB and other forest biophysical parameter measured from field data with PALSAR 50m FBD data. σ^0 HV and HH/HV shows good correlation with forest biomass. Then, a multi-linear regression model approach has been used. Our results showed that most of the Cambodian forests (52%) falls into the 100-200 Mg/ha biomass value. About, 28% of Cambodian forests falls into biomass class >200 Mg/ha. The total biomass in evergreen and deciduous forests show good synergies with the Sato, (2011) estimated biomass using 100 PSPs, although the methodological approaches are different. Such national level biomass map is not very precise and accurate but it can provide general information about biomass distribution which is needed for forest management practices in a cost effective way. PALSAR 50m mosaic data shows saturation at about 150-200 Mg/ha. The saturation problem can be overcome using polarimetric-interferometry SAR (PolInSAR) technique or P-band SAR data. For more precise estimation we must look forward for the P-band SAR or DESDynl satellite system in the future. Further research may focus on the integration of multi-source data, which involves the effective integration of remote sensing, GIS and modeling techniques. The limitations of the applicability of this methodology to other regions may differ with forests types because of the difference in tree species as well as in canopy structure with different environmental conditions. In case of Cambodia most of the forests (75%) are in flat area therefore the topographic effects of PALSAR is less as compared to other tropical countries.

Chapter 7

ROLE OF REMOTE SENSING AND COMMUNITY FORESTRY TO IMPLEMENT REDD+ POLICIES

7.1. INTRODUCTION

Forests are one of the greatest natural assets which, provide ecological, social and economic services. They act as a sink for global carbon cycle (Running and Nemani, 1988; Running et al., 1989; Sivanpillai et al., 2006). Ecologically, forests are habitat for many animals and plants species and they play a key role in nutrient cycling, hydrology and other vital ecosystem functions. Forests are economically important to humans and they are used for timber, building materials, paper, fuel and other requirements. They also provides livelihood to the local and indigenous people. To meet the growing demands of forest products globally for rapid developments, the depletion of forest resources have been accelerating since last few decades (Suzuki et al., 2006). Khatun (2011) has noticed that the major strategies to decrease atmospheric CO_2 through preserving existing forest carbon stocks and planting trees by better management techniques. Therefore, adoption of appropriate management practices is required to ensure ecological integrity and long-term sustainability of forest resources.

To mitigate climate change, most of the present research is being concentrated on afforestation, reforestation, reducing deforestation and degradation to minimize atmospheric CO_2 levels. This can be accomplished by examining the present forest management plans of developing countries and their strength, weakness and opportunities in order to effective implementation of REDD+ policies. REDD+ has given high priority to mitigate climate change in last COP 16 (Conference of Parties). It will not only provide financial support to the developing countries but also provide financial benefit to the communities and indigenous people (Angelson, 2009; Costenbader, 2011). Local and indigenous people can play an important role to protect forest and other ecosystem services because they have adequate knowledge of ground based reality.

Most of the forest management plans are traditionally based on production in even-aged forest stands, with planting, thinning, and final felling (Backeus, 2009). Previous studies have shown that prolonging the rotation period or reducing the intensity of the thinning can result in increasing carbon sequestration (Kaipainen et al., 2004; Liski et al., 2001; Kellomäki and Leinonen, 2005; Pohjola et al., 2004). Conservation of existing forest cover is crucial for success of future REDD+ strategies to mitigate climate change. It will be possible by controlling the drivers of deforestation. Hence, updated information about forest cover and deforestation will be helpful for the prediction of deforestation as well as selection of suitable sites for thinning and plantation practices.

Advanced remote sensing techniques can be effectively used to map forest cover and deforestation. However, Measurement of forest biomass still has some uncertainties. These uncertainties are mainly because of significant errors in sampling plots locations, tree biophysical parameters (DBH, height), allometric models, saturation problem, geometric and radiometric corrections, and modeling of the relationship of AGB with spectral response. The key to reduce uncertainties in these parameters is to identify their sources and minimize them (Wang et al., 2011). In order to minimize these uncertainties in biomass measurement, the participation of communities' can certainly help.

This chapter is focused on the elaborating the application and limitation of remote sensing techniques for the management of Cambodian forests as well as encouraging the role of communities' for forest management. In this context, it is important to obtain reliable and consistent information of forest cover, deforestation, and forest biomass to support sustainable forest management. The results from this chapter will hopefully provide guidance for decision-makers as well as other researchers regarding the integrated role of remote sensing and community people in relation to sustainable forest management.

7.2. Study Area

Cambodia has a population of about 13.4 million, of which 81% lives in rural areas (NIS, 2008). Cambodia's population has increased by 1.95 million with an annual growth rate of 1.5% during the last decades (Ra et al., 2011). Most of the rural population lives in traditional woody houses and depends on agriculture and forestry resources. In 2008, forestry sector has contributed about 7% to GDP (Chao, 2009). Fuel-wood, foods, traditional medicines, rattan, resins, and construction materials are the main products to the local people. Forests also provide food security, employment, health maintenance and household income to the local people (McKenney and Tola, 2002). Hansen and Top (2006), in a study of livelihoods in 16 villages, reported that poor and medium households obtained 42% and 30% of their annual income from forest resources respectively. These findings demonstrate that forest products play a critical role in supporting rural livelihoods in Cambodia.

During the last decade, rapid population growth and economic development have led the country's forests under huge pressure. The major causes of deforestation in Cambodia are illegal logging, forestland conversion, heavy reliance on fuel wood for energy, lack of capacity for sustainable forest management and financial incentive for forest conservation. In addition, Economic Land Concessions (ELCs) and insecure land tenure are also among the major drivers of deforestation in the country (Fox et al, 2008; Poffenberger, 2009; Van Beukering, 2009; UNEP, 2009).

Table 7.1 represents the area under various types of forest protection system of Cambodia. Economic land concession covers about 8.8% of total forest area, however community forestry area covers only 3% of Cambodia's total forest area. For sustainable management of forest resources, Cambodian government should promote community forestry programme on large scale.

Table 7.1. Statistics of Cambodian protected area system

No.	Name	Designation	Area (ha)	Area (ha)	Administrative Responsibility
1	Kulen Promtep	Wildlife Sanctuary	409600	1916100	Ministry of Environment
2	Lomphat		252600		
3	Boeng Per		249700		
4	Phnom Prich		224200		
5	Phnom Nam Lyr		54900		
6	Phnom Samkos		329700		
7	Phnom Aural		254200		

Table 7.1. (Continued)

No.	Name	Designation	Area (ha)	Area (ha)	Administrative Responsibility
8	Snoul		75300		
9	Roniem Daun Sam		40000		
10	Peam Krasop		25900		
11	Virachey	National Park	334200	759700	
12	Phnom Kulen		37400		
13	Kirirom		33600		
14	Bokor		149900		
15	Ream		18400		
16	Botum Sakor		183400		
17	Kep		2800		
18	Samlaut	Multiple Use Area	61000	421000	
19	Dong Peng		28600		
20	Tonle Sap		331400		
21	Banteay Chhmar	Protected Landscape	84800	101800	
22	Angkor		13700		
23	Preah Vihear		3300		
24	Stung Treng	Ramsar Site	14900	14900	
25	Southern Cardamoms	Protected Forest	145100	1586900	Ministry of Agriculture, Forestry and Fisheries
26	Preah Vihear		190000		
27	O Ya Dao		100000		
28	Ang Trapeng Thmor		12900		
29	Mondulkiri		430700		
30	Central Cardamoms		400700		
31	Phnom Thamu		2400		
32	Kbal Chhay		6400		
33	Seima Biodiversity conservation area		298700		
34	Koh Ker Historical Site	Cultural Site	7900	7900	Ministry of Cultural and Fine Arts
	Total		**4808300**	**4808300**	

Source: BPAMP, 2007.

7.3. Discussion

7.3.1. Forest Management Strategies and Carbon Sequestration

Management of forest resources is a crucial factor to mitigate the effects of climate change. According to Bravo et al., (2008) forest management is possible by a number of strategies, including: (a) conservation and maintenance of existing forest carbon stocks (b) increasing carbon stocks through afforestation and reforestation (c) modification of the forest species composition and tree size distributions (d) promoting the planting of more resilient tree genotypes (e) planting trees to stabilize soils and to reduce the expected impacts of rainfall and temperature changes (f) regular thinning to restore forest and accelerate carbon sequestration (Dwyer et al., 2010).

Fire protection, pest control, increasing the length of time to rotation, regulation of tree density, improvement of the nutritional state, and management of residues are another various types of management options that may increase the forest carbon stock as well as ecosystem services and biodiversity (Bravo et al., 2008). Forest age can give information about rotation length because at old age, forest carbon sequestration decreases slightly or may continue to increase slowly. Therefore, knowledge of appropriate rotation period is needed for the natural regeneration of young plant canopies (Paul et al., 2002). Reforestation and effective conservation of forest area could lead to the carbon sequestration and biodiversity conservation (Nabuurs et al., 2007). Promotion of natural regeneration in disturbed forests is also simple and low cost forest restoration method (Shono et al., 2007).

Collection of basic information about forests parameters is necessary in order to implement forest management practices. Information about forest cover, deforestation, degradation, and forest biomass is required for making appropriate sustainable forest management plan. These informations could be generated using remote sensing techniques at national level as well as forest inventory data at local level. Satellite data is useful to monitor forest periodically at national level and the role of community people is important to collect forest inventory parameters at local level.

7.3.2. Use of Remote Sensing Techniques to Monitor Forests

Remote sensing techniques have played a crucial role to study forest cover, deforestation and forest biomass on a spatio-temporal scale.

Development of remote sensing techniques with the application of optical SAR (Synthetic Aperture Radar) and LiDAR (Light detection and ranging) data have made mapping of forest parameters cost and time effective with significant accuracy. Most of the present monitoring systems are based on optical and SAR data. Remote sensing techniques are feasible to study about forest environmental conditions (topography, slope, soil type, soil moisture etc.), and zoning of forests under various environmental conditions. Forest environment condition map can be used to make appropriate forest management practices for example forest located on a mountainous area with good supply of water have less human induced logging and such sites should be rich of forest carbon and biodiversity. Remote sensing can also provide information about dense, sparse, young and old types of forests. Lal and Singh (2000) has noticed that high carbon sequestration rates in young forests and decline in older age forest. Therefore, information about forest density and age supplied by remote sensing can be used for regular thinning to maintain high carbon stocks in forests.

Updated information about forest cover, deforestation and forest biomass can be used to identify the deforested sites and prediction of deforestation derivers. Figure 7.1 shows the updated forest cover map of Cambodia based on prediction of deforested sites using PALSAR and Landsat data. National level biomass map of Cambodia (Figure 7.2) has been generated based on PALSAR 50m mosaic data (chapter 8). It shows saturation at around 150-200 Mg/ha of biomass because of saturation of PALSAR backscattering properties. However, this biomass map can provide information about high, medium and low density biomass and can be used by foresters to minimize illegal logging in the high biomass region by increasing the patrolling of such sites. The forests with low biomass can be used for the reforestation sites to increase the biomass. Selection of degraded and unproductive land using remote sensing techniques can be used for afforestation to increase the forest cover. Application of GIS techniques for the selection of sites for reforestation, afforestation, agriculture expansion and community forestry projects is also based on updated information about forest cover, deforestation and forest biomass. Measurement of forest biomass using remote sensing techniques has few limitations and the recent studies (Skutsch, 2010; Danielsen et al., 2011; Fry, 2011; Pratihast and Herold, 2011) have suggested that involvement of community people could help to overcome these limitations. Table 7.2 shows the comparison of remote sensing and community based monitoring of forests. It shows that community based monitoring can provide more parameters with high accuracy, however cover of large and remote areas is not possible by community based monitoring. Whereas remote sensing can provide information on large scale on spatial and temporal scale.

Table 7.2. Comparison of remote sensing and community based monitoring of forests (Olander et al., 2008; Pratihast and Herold, 2011)

No.		Cost	Forest inventory parameters				Deforestation	Degradation	Forest cover
			DBH	Height	Species	density			
1	Coarse resolution (AVHRR, MODIS, SPOT-VEGETATION)	Free	-	-	-	-	+	-	+
2	Medium resolution (Landsat, SPOT, ASTER, AVNIR, IRS, CBERS)	Free to low	-	-	-	-	++	+	++
3	Fine resolution (IKONOS, Quickbird)	high	-	-	-	++	+++	+++	+++
4	SAR (Radarsat, PALSAR, ENVISAT, TerraSAR-X)	Free to high	-	+	-	+	+++	+	+++
5	LiDAR	High	-	+++	-	++	+++	+++	+++
6	Community based monitoring	high	+++	+++	+++	+++	+++	+++	++

Quality indicator: (-) no, (+) low, (++) medium, (+++) high

7.3.3. Role of Community Forestry to Monitor Forests

Community based forest monitoring and conservation has been proposed an additional and effective way to overcome limitations of remote sensing and increase the reliability of forest monitoring in cost effective way (Danielsen et al., 2011; Pratihast and Herold, 2011; Fry, 2011). Engagement of local communities in forest monitoring can be advantageous because: (a) local communities have in-depth knowledge of local forest and forest species (b) local communities have easy access to the surrounding and can make regular field visits (c) local communities have information about probable causes of deforestation and forest degradation so they can minimize them (d) local communities can do patrolling to conserve forest from illegal loggers (e) active involvement of communities can promote long term sustainability (Pratihast and Herold, 2011). Table 7.3 shows the pros and cons of the community forestry monitoring.

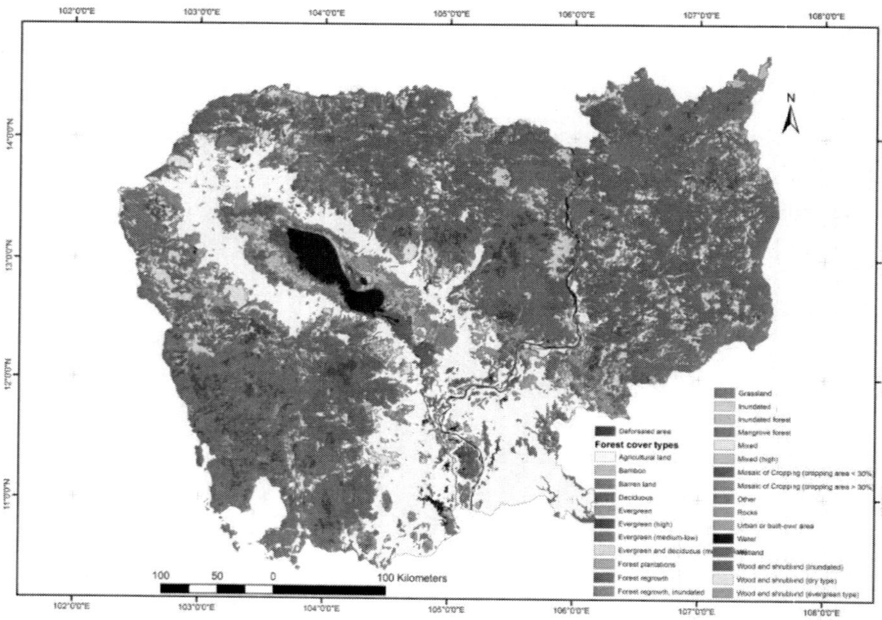

Figure 7.1. Updated forest cover map of Cambodia (year 2009).

Previous studies noticed that promotion of community forestry projects will improve livelihood of indigenous people as well as poverty alleviation. A study by Bray et al. (2008) showed that community forest might generate more incomes for local people than protected areas. In Cambodia most of the community forestry project sites are focused on degraded forests (Poffenberger, 2006). These community projects work on restoration of forests by reforestation activities.

Skutch et al., (2011) has compared the cost of community based forest carbon stock monitoring and expert based monitoring and results showed that expert based monitoring costs 2-3 times of community based monitoring. Expert based monitoring is more mostly because of the higher expenditure in their air travel, local travel, logistics and salaries to experts (Balmford et al.,

2003). Fry, (2011) has also noticed that community based monitoring is feasible, reliable and cheaper than expert based monitoring.

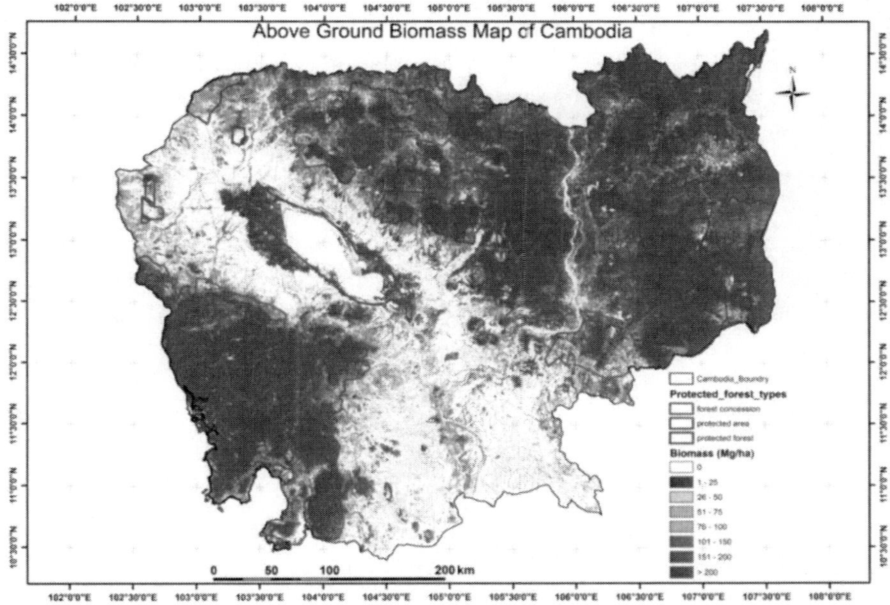

Figure 7.2. Biomass map of Cambodia overlaid with forest protection types.

Table 7.3. Pros and cons of community based forest monitoring (Larrazabal and Skutch, 2011)

No.	Pros	Cons
1	Large number of local people can be employed to facilitate collection of large amount of data within their local area so it will also control urban immigration.	In the beginning training and capacity building will be needed to ensure accuracy and reliability of data
2	Local people can provide important ecological and biodiversity data because of in-depth knowledge about the local area	In the early stage proper supervision may be required to ensure good quality of good
3	Cost will be low as compare to survey by experts	Reliability of community measured data would have to be assured
4	Community will be able to carry out repeated monitoring because of closeness to the project site	Local people might be tempted to exaggerate the carbon stock increases if they are rewarded on the basis of carbon stock
5	Local people can get local livelihood	Independent verification will be required

In a nutshell, remote sensing techniques are the main tools used at the national level monitoring of forest. However, local level community data can also be an additional input to monitor forests. Thus, integrated approach to link the national level forest information by remote sensing and local communities based monitoring would be win-win situation. Operational use of this integrated approach of remote sensing and community based monitoring needs capacity building. Seminars, workshops and short/long-term training courses for capacity building are necessary to improve the capability of community people as well as government level staff in order to implement sustainable forest management plan. Information about forest cover, topography, roads, canals, aerial photograph, satellite data, forest biomass map etc. can make foresters and community people to coordinate forest management plans.

Conclusion

Forest resources are of great importance and of immense value to mankind in the present and in the future. They are being degraded at an alarming rate by various activities. Periodical monitoring of forest resources are important for better management plans. The satellite based sensors provide valuable information useful in assessment, monitoring and management of forest ecosystems. This paper demonstrated the use of forest cover, deforestation and forest biomass information for making an effective sustainable forest management plan. This chapter focused on the site suitability for the planned thinning, reforestation, allocation of forests for community forestry and concession companies. Integrated approach of remote sensing and community based monitoring can be a vital data source for REDD+ policies implementation. That will be a good option for forests management and forests conservation because it involves various stakeholders from government to community level. This approach will be useful for making national level forests management plans and policies.

Chapter 8

SUMMARY AND RECOMMENDATIONS

No single sensor or any satellite system whether optical, SAR or LiDAR can be expected to provide consistent information about forest cover, deforestation and forest biomass. Therefore, the use of multi-sensor remote sensing techniques can potentially overcome the limitations of each sensor. This book has explored the potential of multi-sensor remote sensing techniques for understanding forest cover, deforestation, and forest biomass.

The overarching question was how to use remote sensing techniques, especially multi-sensor remote sensing techniques, for technological development of the REDD+ MRV system. Methodologies for forest cover, deforestation and AGB estimation has been developed to study Cambodian forests for effective implementation of REDD+ mechanism and forest management plans. It has been considered that all the chapters build on each other to achieve the goal as illustrated in Figure 1.3.

To achieve the goal of this book, the work on the development of the REDD+ MRV system was tried, which is still under development. Various scientific communities and Subsidiary Body for Scientific and Technological Advice (SBSTA) are working on the development of these methodologies. The proposed methodology is an initiative towards MRV development but is still not perfect because of some limitations. Therefore, further research is needed for more refinement of the methodologies for effective implementation of REDD+ mechanism. Chapter 6 provides national level biomass mapping that is needed for national level REDD+ strategies development and for patterns of forest carbon stocks in various types of forests of Cambodia. Chapter 7 provides use of forest cover and forest biomass information for the implementation of national level sustainable forest management plan.

8.1. Chapter 3: Forest Cover Monitoring Based on Full Polarimetric PALSAR Data

This chapter demonstrates the potential of full polarimetric L-band PALSAR data for forest cover classification. Cloude-Pottier H/A/Alpha, Freeman-Durden three component and Yamaguchi four component decomposition were used to study the forest cover in Cambodia. Application of polarimetric decomposition techniques for forest cover classification to characterize various targets was based on physical, dielectric and scattering properties. Most of the previous studies have not shown the best SAR based decomposition. Therefore, it has been carried out in this study. A quantitative comparison has been made among various decomposition techniques to find the best SAR based decomposition classification for forest cover monitoring and to understand the potential of full polarimetric PALSAR data to generate forest cover maps. This chapter shows an overestimation of volume scattering in Freeman-Durden three component decomposition. Yamaguchi four component decomposition based classification shows an overall accuracy of 94.1%. This chapter is useful for detection of structural differences between the forests canopies, which will be useful for forest management practices to implement climate change mitigation policies like REDD+. The new generation of full polarimetric SAR technology could help to study forest parameters for better implementation of REDD+ mechanism.

Further work: Further scientific research in this area includes evaluation of the classification accuracy. Use of high resolution SAR data such as TerraSAR-X and combination of X-band and L-band data for forest cover classification is also a future scope for getting more precise information about forest cover types.

8.2. Chapter 4: Forests and Deforestation Characterization Using PALSAR Data

This chapter demonstrates the capability of full polarimetric PALSAR data for characterization of the forests and deforestation in Cambodia, to support REDD+ climate change mitigation policies. Mean backscattering coefficient (σ^o), entropy (H), alpha angle (α), anisotropy (A), pedestal height (PH), radar vegetation index (RVI), Freeman-Durden three-component decomposition and Yamaguchi four-component decomposition parameters have been observed for various forests types. The observations show that the forest-types and deforested areas are having variable polarimetric and backscattering properties because of the structural difference. Evergreen forests are characterized by a high value of σ^o HV (-12.96 dB) as compared with deforested areas (σ^o HV = -22.2 dB). The values of polarimetric parameters such as entropy (0.93), RVI (0.91), pedestal height (0.41) and Freeman-Durden volume scattering (0.43) are high for evergreen forests, whereas deforested areas have low values of entropy (0.36) and RVI (0.17). The observed entropy and RVI were highest for evergreen forests and lowest for deforested areas, whereas the pattern was reverse for HH/HV polarization ratio. The observation shows that σ^o HV, polarization ratio (HH /HV), H, RVI, PH and Yamaguchi four component based decomposition give better results for the characterization of forest types and deforested areas when compared with other polarimetric parameters.

Further work: Integration of optical data with the PALSAR polarimetric features can be useful for monitoring deforested areas because the repeat cycle of optical data is more than PALSAR data. The PALSAR data is not affected by clouds as compared to optical sensor data.

8.3. Chapter 5: Deforested Area Height Estimation Using DEMs Data

Deforestation has been a major cause of climate change and other environmental problems. An accurate estimation of the volume of deforested area is needed for calculation of emission factor (EF) for REDD+. This chapter demonstrated the estimation of height of deforested area using different digital elevation models (DEMs) in Cambodia. The results obtained from the ICESat-GLAS, PRISM-DSM, ASTER-GDEM and STRM DEM clearly showed the height of the deforested area. The height of cut over forest generated from the PRISM-DSM and SRTM-DEM is more reliable than that from the PRISM-DSM and GDEM.

Further work: Estimation of forest height with combination of P-band and X-band SAR data is also future scope. Application of PolInSAR techniques to estimate tree height without temporal and volume decorrelaion is also future scope.

8.4. Chapter 6: PALSAR 50m Mosaic Data Based National Level Biomass Estimation

PALSAR 50m mosaic data has been used that is freely available at Japan Aerospace Exploration Agency (JAXA) and forest inventory data to estimate national level biomass. This chapter investigated the potential of PALSAR FBD 50m mosaic data to estimate above ground biomass (AGB) in Cambodia. AGB was estimated using a bottom-up approach based on field calculated biomass and backscattering (σ^o) properties of PALSAR data. The relationship between the PALSAR σ^o HV and HH/HV with field based biomass was strong with $R^2 = 0.67$ and 0.56, respectively. PALSAR estimated biomass shows good results in deciduous forests because of less saturation compared to dense evergreen forests. The validation result shows high coefficient of determination $R^2 = 0.61$ with RMSE = 21 t/ha using values up to 200 t/ha biomass. There are some uncertainties because of the field-based measurement and saturation of PALSAR data.

Further work: At high biomass regions the PALSAR become saturated, hence such biomass maps are not very precise, therefore it is required to think about how to overcome the saturation problem of PALSAR data. For more precise estimation we must look forward for the P-band SAR or DESDynl satellite system in the future.

8.5. Chapter 7: Role of Remote Sensing and Community Forestry to Manage Forests

This chapter theoretically covers the present forest management plan of the Cambodian Forestry Administration and their readiness for the implementation of REDD+ mechanism. It also demonstrates the use of updated forest cover and forest biomass information for making an effective sustainable forest management plan. This chapter focused on the site suitability for the planned thinning, reforestation and allocation of forests for community forestry. Integrated approach of remote sensing and community forestry is useful for sustainable forest management.

Future work: Strategies and methodologies should be further developed for forest and biodiversity conservation. The AGB map will be utilized to provide information about the successful implementation of forest management practices for the REDD+ assessment and policies implementation at a national level.

In summary, the chapters presented in this book contribute towards methodological development for MRV establishment as well as a technical part of future REDD+ mitigation scheme. The capacity-building of local people and government is necessary for the success of REDD+ projects, hence study of socio-economic aspects of REDD+ is also important.

References

Achard F., Defries R., Eva H., Hansen M., Mayaux P., Stibig H. J., 2007. Pan-Tropical monitoring of deforestation. *Environment Research Letters*, 2, 045022, 1-11.

Achard F., Stibig H., Eva H. D., Lindquist E. J., Bouvet A., Arino O., Mayaux P., 2010. Estimating tropical deforestation form earth observation data. *Carbon Management,* 1, 2, 271-287.

Ahren F. J., Drdle T., Maclean D. A., Kneppeck I. D., 1991. A quantitative relationship between forest growth rates and thematic mapper reflectance measurements. *International Journal of Remote Sensing*, 12, 387-400.

Ainsworth, T. L., Kelly, J. P., Lee, J. S., 2009. Classification comparisons between dual-pol, compact polarimetric and quad-pol SAR imagery. *ISPRS Journal of Photogrammetry and Remote Sensing,* 64, 464-471.

Alberga V., 2007. A study of land cover classification using polarimetric SAR parameters. *International Journal of Remote Sensing*, 28, 17, 3851-3870.

Amariei L., 2004. Capacity building for law compliance in the forest sector case study: Cambodia. Final report prepared for FAO/ITTO, 1-30.

An W., Xie C., Yuan X., Cui Y., Yang J., 2011. Four-component decomposition of polarimetric SAR images with deorientation. *IEEE Transactions on Geoscience and Remote Sensing,* doi: 10.1109/LGRS.2011.2157078.

Anderson M. C., Neale C. M. U., Li F., Norman J. M., Kustas W. P., Jayanthi H., Chavez J., 2004. Upscaling ground observations of vegetation water content, canopy height, and leaf area index during SMEX02 using aircraft and Landsat imagery. *Remote Sensing of Environment,* 92, 447–464.

Angelsen A., 2008. Moving ahead with REDD Issues, options and implications, CIFOR, Indonesia.

Angelsen A., 2009. Realising REDD+ national strategy and policy options. CIFOR, Indonesia.

Angelsen A., Brown S., Loisel C., Peskett L., Streck C., Zarin D., 2009. Reducing Emissions from Deforestation and Forest Degradation (REDD): an options assessment report. Washington, DC: Meridian Institute.

Ashwall D., Ogonowski M., Neou S., McCulloch C., 2011. Assisting Cambodian policymakers with designing REDD plus approaches under a post-2012 international climate change policy framework. *Center for Clean Air Policy,* 45-47.

Asner P. G., 2009. Tropical forest carbon assessment: integrating satellite and airborne mapping approaches. *Environment Research Letter,* 4, 1-11.

Asner G. P., Powell G. V. N., Mascaro J., Knapp D. E., Clark J. K., Jacobson J., Kennedy-Bowdoin T., Balaji A., Paez-Acosta G., Victoria E., Secada L., Valqui M., Hughes R. F., 2010. High-resolution forest carbon stocks and emissions in the Amazon. PNAS, doi/10.1073/pnas.1004875107.

Austin J. M., Mackey B. G., Van Niel K. P., 2003. Estimating forest biomass using satellite radar: an exploratory study in a temperate Australian Eucalyptus forest. *Forest Ecology and Management*, 176, 1-3, 575-583.

Avtar R., W. Takeuchi, H. Sawada, 2011a. Full Polarimetric PALSAR based Land Cover Monitoring in Cambodia for Implementation of REDD Policies. *International Journal of Digital Earth*. DOI:10.1080/ 17538947. 2011.620639.

Avtar R., H. Sawada, W. Takeuchi, G. Singh 2011b. Characterization of Forests and Deforestation in Cambodia Using ALOS/PALSAR Observation. *Geocarto International.* DOI:10.1080/10106049.2011. 626081.

Avtar R., W. Takeuchi, H. Sawada, 2011c. Assessment of Cashew and Rubber plants biophysical parameters based on ALOS/PALSAR data. Seisan-Kenkyu, 63, 4, 51-54.

Backeus S., 2009. Forest management strategies for CO_2 mitigation. Doctoral thesis, Swedish University of Agricultural Sciences.

Balmford, A., Green, R.E. & Jenkins, M., 2003. Measuring the changing state of nature. *Trends in Ecology & Evolution,* 18, 7, 326-330.

Beaudoin A., Le Toan T., Goze S., Nezry E., Lopes A., Mougin E., 1994. Retrieval of forest biomass from SAR data. *International Journal of Remote Sensing,* 15, 4, 2777-2796.

Bhang K. J., Schwartz F. W., Braun A., 2007. Verification of the vertical error in C-band SRTM-DEM using ICESat and Landsat-7, Otter Tail County,

MN. *IEEE Transactions on Geoscience and Remote Sensing,* 45, 1, 36–44.

Blaser J., Sarre A., Poore D., Johnson S., 2011. Status of Tropical Forest Management 2011. IT TO Technical Series No 38. International Tropical Timber Organization, Yokohama, Japan, 146-157.

Burgess N.D., Bahane B., Clairs T., 2010. Getting ready for REDD + in Tanzania: a case study of progress and challenges. *Oryx,* 44, 339–351.

BPAMP. 2007. Cambodia's National Protected Area System: A Gap Analysis of its Representativeness and Effectiveness. Biodiversity and Protected Areas Management Project, Phnom Penh, Ministry of Environment Cambodia.

Bravo E., LeMay V., Jandl R., Gadow K. V., 2008. Managing forest ecosystems: The challenges of climate change. *Springer publication,* 179-192.

Bray, D. B., E. Duran, V. H. Ramos, J.-F. Mas, A. Velazquez, R. B. McNab, D. Barry, and J. Radachowsky. 2008. Tropical deforestation, community forests, and protected areas in the Maya Forest. *Ecology and Society* 13, 2, 56.

Breugel M. V., Ransijin J., Craven D., Bongers F., Hall J. S., 2011. Estimating carbon stock in secondary forests: Decisions and uncertainties associated with allometric biomass models. *Forest Ecology and Management,* doi:10.1016/j.foreco.2011.07.018.

Brown S., Gaston G., 1995. Use of forest inventories and geographic information systems to estimate biomass density of tropical forests: Application to tropical Africa. *Environmental Monitoring and Assessment,* 38, 2-3, 157-168.

Brown S., 1997. Estimating biomass and biomass change of tropical forests: A primer. *FAO Forestry Paper* 134, Rome, 55.

Brown S., 2002. Measuring carbon in forests: current status and future challenges. *Environmental Pollution,* 116, 363–72.

Bunker D. E., DeClerck F., Bradford J. C., Colwell R. K., Perfecto I., Phillips O. L., Sankaran M., Naeem S., 2005. Species loss and aboveground carbon storage in a tropical forest. *Science,* 310, 1029–1031.

Butler R., 2006. Calculating the value of avoided deforestation. Retrieved from: http://news.mongabay.com/2006/1031-deforestation.html.

Carle J., Holmgre P., 2008. Wood from planted forests: A global outlook 20052030. *Forest Products Journal,* 58, 12, 6-18.

Castro K., Sanchez-Azofe ifa G., Rivard B., 2003. Monitoring secondary tropical forests using space-borne data: Implications for Central America. *International Journal of Remote Sensing,* 24, 1853 – 1894.

Chao, L. 2009. Agricultural mechanization and agricultural development strategies in Cambodia. Country report for fifth session of the technical committee of UNAPCAM.14-15 October 2009, Los Banos.

Chen qi., 2010. Assessment of terrain elevation derived from satellite laser altimetry over mountainous forest areas using airborne lidar data. *ISPRS Journal of Photogrammetry and Remote Sensing,* 65, 1, 111-122.

Clark M. L., Clark D. B., Roberts D. A., 2004. Small-footprint lidar estimation of subcanopy elevation and tree height in a tropical rain forest landscape. *Remote Sensing of Environment,* 91, 1, 68–89.

Claudia C. C., Harding D. J., 2006. SRTM C-band and ICESat laser altimetry elevation comparisons as a function of tree cover and relief. *Photogrammetric Engineering & Remote Sensing,* 72, 3, 287-298.

Cloude S. R., Pottier E., 1996. A review of target decomposition theorms in radar polarimetry. *IEEE transactions on Geoscience and Remote Sensing,* 34, 2, 498-518.

Cloude S.R., Pottier E., 1997. An entropy based classification scheme for land applications of polarimetric SAR. *IEEE Transactions on Geoscience and Remote Sensing,* 35, 1, 68-78.

Cohen W. B., Spies T. A., Fiorella M., 1995. Estimating the age and structure of forest in a multi-ownership landscape of western Oregon. *International Journal of Remote Sensing,* 16, 721-746.

Cohn S. A., Mayor S. D., Grund C. J., Weckwerth T. M., Senff C., 1998. The Lidars in flat terrain (LIFT) experiment. *Bulletin of American Meteorological Society,* 79, 7, 1329-1343.

Collins J. N., Hutley L. B., Williams R. J., Boggs G., Bell D., Bartolo R., 2009. Estimating land-scape vegetation carbon stocks using airborne multi-frequency polarimetric synthetic aperture radar (SAR) in the savannahs of north Australia. *International Journal of Remote Sensing,* 30, 5, 1141-1159.

Congalton R. G., 1991. A review of assessing the accuracy of classification of remote sensing data. *Remote Sensing of Environment,* 37, 35-46.

Coops N. C., 2002. Eucalypt forest structure and synthetic aperture radar backscatter: a theoretical analysis. *Trees,* 16, 28-46.

Costenbader J., 2011. REDD+ benefit sharing: A comparative assessment of three national policy approaches. *Forest Carbon Partnership,* 33-49.

Cotula L., Mayers J., 2009. Tenure in REDD – Start-point or afterthought? Natural Resource Issues No. 15. International Institute for Environment and Development. London, UK.

Crocker C. D., 1962. The soils of Cambodia, exploratory survey. Royal Cambodian Government Soil Commission and USAID Joint Publication, Phnom Penh, 83.

Danielsen, F., Skutsch, M., Burgess, N. D., Jensen, P. M., Andrianandrasana, H., Karky, B., Lewis, R., Lovett, J. C., Massao, J., Ngaga, Y., Phartiyal, P., Poulsen, M. K., Singh, S. P., Solis, S., Sorensen, M., Tewari, A., Young, R. and Zahabu, E., 2011. At the heart of REDD+: a role for local people in monitoring forests?. *Conservation Letters,* 4, 2, 158-167.

Defries R., Achard F., Brown S., Herold M., Murdiyarso D., Schlamadinger B., De Souza Jr C., 2007. Earth observations for estimating greenhouse gas emissions from deforestation in developing countries, *Environmental science & policy,* 10, 385-394.

Dixon R. K., Brown S., Houghton R. A., Solomon A. M., Trexler M. C., Wisniewski J., 1994. Carbon pools and flux of global forest ecosystems. *Science* 263, 185-190.

Djomo A. N., Knohl A., Gravenhorst G., 2011. Estimations of total ecosystem carbon pools distribution and carbon biomass current annual increment of a moist tropical forest. *Forest Ecology Management,* doi:10.1016/j.foreco.2011.01.031.

Dobson M. C., Ulaby F. T., Le Toan T., Beaudoin A., Kasischke E. S., Christensen N., 1992. Dependence of radar backscatter on coniferous forest biomass. *IEEE Transaction on Geoscience and Remote Sensing,* 30, 412–415.

Dobson M. C., Ulaby F. T., Pierce L.E., Sharik T. L., Bergen K. M., Kellndorfer J., Kendra J. R., Li E., Lin Y.C., Nashashibi A., Sarabandi K. L., Siqueira P., 1995. Estimation of Forest Biophysical Characteristics in Northem Michigan with SIR-C/X-SAR. *IEEE Transactions on Geoscience and Remote Sensing,* 33, 4, 877-896.

Dong J., Kaufmann R. K., Myneni B. R., Tucker J. C., Kauppi E. P., liski J., Buermann W., Alexeyev V., Hughes K. M., 2003, Remote sensing estimates of boreal and temperate forest woody biomass: carbon pools, sources, and sinks, *Remote sensing of environment,* 84, 393-410.

Drake J. B., Dubayah R. O., Clark D. B., Knox R. G., Blair J. B., Hofton M. A., Chazdon R. L., Weishampel J. F., Prince S. D., 2002. Estimation of tropical forest structural characteristics using large-footprint lidar. *Remote Sensing of Environment,* 79, 305–319.

Drake J. B., Knox R. G., Dubayah R. O., Clark D. B., Condit R., Blair J. B., Hofton M. A., 2003. Above-ground biomass estimation in closed canopy neotropical forests using lidar remote sensing: Factors affecting the generality of relationships. *Global Ecology and Biogeography,* 12, 147–159.

Drezet P. M. L., Quegan S., 2007. Satellite-bsed radar mapping of British forest age and Net Ecosystem Exchange using ERS tandem coherence. *Forest Ecology and Management,* 238, 65-80.

Dudley R. G., 2010. A little REDD model to quickly compare possible baseline and policy scenarios for reducing emissions from deforestation and forest degradation. *Mitigation and Adaptation Strategies for Global Change* 15, 1, 53-69.

Dwyer J. M., Fensham R., Buckley Y. M., 2010. Restoration thinning accelerates structural development and carbon sequestration in an endangered Australian ecosystem. *Journal of Applied Ecology*, 47, 681–691.

Economic land concession, MAFF, 2010. http://www.elc.maff.gov.kh/en/news/12-elc-status.html (access on 19/09/2011).

Englhart S., Keuck V., Siegert F., 2011. Aboveground biomass retrieval in tropical forests – the potential of combined X- and L-band SAT data use. *Remote Sensing of Environment,* doi:10.1016/j.rse.2011.01.008.

Erikson L. E. B., Santoro M., Fransson J. E. S., 2008. Temporal decorrelation for forested areas observed in spaceborne L-band SAR interferometry. IGARSS, 2008.

Evans J., Turnbull J., 2004. Plantation Forestry in the Tropics (3rd ed.), Oxford University Press, Oxford.

Evans J., 2009. Forests uses, impacts & sustainability, FAO, MPG Books group, UK.

F. A., 2006. Community forestry statistics.

F. A., 2009. Community forestry statistics.

Fang C., Wen H., 2005. A new classification method based on Cloude Pottier eigenvalue/eigenvector decomposition. IGARSS, 2005.

Fang C., Wen H., Yirong W., 2006. An improved Cloude-Pottier decomposition using H/α/SPAN and complex Wishart classifier for polarimetric SAR classification. CIE06, International conference on Radar, Shanghai.

Fang J. Y., Wang G. G., Liu G. H., Xu S. L., 1998. Forest biomass of China: an estimate based on the biomass–volume relationship. *Ecological Applications,* 8, 4, 1084–1091.

FAO, 2005. Global Forest Resources Assessment, Food and Agriculture Organisation of the United Nations, Rome.

FAO, 2006, Global forest resources assessment 2005: Progress towards sustainable forest management, FAO Forestry Paper 147. FAO, Rome, Italy.

FAO, 2010. Global forest resources assessment 2010 country report: Cambodia. http://www.fao.org/docrep/013/al470E/al470E.pdf (access on: 06/03/2011).

FAO, 2010. Planted forests in sustainable forest management: A statement of principles.

FAO, 2011. Media Centre, Official webpage of FAO. http://www.fao.org/news/story/pt/item/40893/icode/en/ (Accessed on 06-03-2011)

Fassnacht K. S., Gower S. T., MacKenzie M. D., Nordheim E. V., Lillesand T. M., 1997. Estimating the leaf area index of North Central Wisconsin forests using the landsat thematic mapper. *Remote Sensing of Environment,* 61, 229–245.

Ferro-Famil L., Pottier E., Lee J. S., 2001. Unsupervised classification of multifrequency and fully polarimetric SAR images based on the H/A/Alpha Wishart classifier. *IEEE Transactions on Geoscience and Remote Sensing,* 39, 11, 2332-2342.

Foody G. M., Boyd D. S., Cutler M. E. J., 2003. Predictive relations of tropical forest biomass from Landsat TM data and their transferability between regions. *Remote Sensing of Environment,* 85, 4, 463–474.

Forest Administration., 2010. Cambodia Forestry Outlook Study. Prepared by the Forest Administration, Cambodia. Asia-Pacific Forestry Sector Outlook Study II. Working Paper No. APFSOS II/WP/2010/32. FAO, Bangkok, Thailand.

Fox, J. M., McMahon, D., Poffenberger, M., Vogler, J., 2008. Land for My Grandchildren: Land Use and Tenure Change in Ratanakiri: 1989-2006. Community Forestry International and the East West Center.

FRA., 2005. Global Forest Resources Assessment, Food and Agriculture Organisation of the United Nations, Rome.

FRA., 2010. Global Forest Resources Assessment, Food and Agriculture Organisation of the United Nations, Rome.

Freeman A., Durden S. L., 1998. A three-component scattering model for polarimetric SAR data. *IEEE transactions on Geoscience and Remote Sensing*, 36, 963–973.

Fricher H. A., Borsa A., Minster B., Carabajal C., Quinn K., Bills B., 2005. Assessment of ICESat performance at the salar de Uyuni, Bolivia. Geophysical Research Letters, 32, L21S06, 5.

Fry, B. F. B., 2011. Community forest monitoring in REDD+: the 'M' in MRV?. *Environmental Science and Policy,* 14, 2, 181-187.

Gama F. F., dos Santos J. R., Mura J. C., 2010. Eucalyptus biomass and volume estimation using interferometric and polarimetric SAR data. *Remote Sensing,* 2, 939–956.

Garcia-Mora T. J., Mas J., Hinkley E. A., 2011. Land cover mapping applications with MODIS: a literature review. *International Journal of Digital Earth,* doi: 10.1080/17538947.2011.565080.

Gardner T. A., Barlow J., Chazdon R., Ewers R. M., Harvey C.A., Peres C. A., Sodhi N. S., 2009. Prospects for tropical forest biodiversity in a human-modified world. *Ecological letters,* 12, 561-582.

Gaughan A. E., Binford M. W., Southworth J., 2009. Tourism, forest conversion, and land transformations in the Angkor basin Cambodia. *Applied Geography*, 29, 212-223.

Gibbs H. K., Brown S., Niles J. O., Foley J. A., 2007a. Monitoring and estimating tropical forest carbon stocks: making REDD a reality. *Environmental Research Letter,* 2, 045023, 13.

Gibbs H. K., Herold M., 2007b. Tropical deforestation and greenhouse gas emissions. *Environmental research letters*, 2, 4, 1-3.

Gillis M. D., 2001. Canada's national forest inventory (Responding to current information needs*). Environmental Monitoring Assessment*, 67, 121-129.

Global Forest Coalition, 2009. REDD Realities: How strategies to reduce emissions from deforestation and forest degradation could impact on biodiversity and Indigenous Peoples in developing countries. Asunción, Paraguay.

Griffiths T., 2007. Seeing 'RED'?: Avoided deforestation' and the rights of Indigenous Peoples and local communities. Forest Peoples Programme. Moreton-in-Marsh, UK.

Guerra J. B., Freitas C. D. C., Mura J. C., 2008. Evaluating potential of band polar data to discrimination increment areas in Amazon rain forest. Proceedings of IEEE IGARSS08.

Hair J. F., Anderson JR R.E., Tatham R.L., Black W.C., 1998. Multivariate data analysis, 5th Ed. Prentice Hall, Upper Saddle River, NJ. 730.

Hajnsek I., Kugler F., Lee S. K., Papathanassiou K.P., 2009. Tropical forest parameter estimation by means of Pol-In-SAR: The INDREX-II campaign. *IEEE Transactions on Geoscience and Remote Sensing*, 47, 2, 481-493.

Hamdan O., Aziz H. K., Rahman K. A., 2011. Remotely sensed L-band SAR data for tropical forest biomass estimation. *Journal of Tropical Forest Science,* 23, 3, 318-327.

Hansen, K. and Top, N. 2006. Economics of Land Use Changes in Cambodia CDRI Cambodia Development Review 10, 2, 6-8.

Harrell P. A., Bourgeau-Chavez L. L., Kasischke E. S., French N. H. F., Christensen N. L., 1995. Sensitivity of ERS-1 and JERS-1 radar data to biomass and stand structure in Alaskan boreal forest. *Remote Sensing of Environment,* 54, 247–253.

Heng S., and Shigeru I., 2002. Community forestry in northern Cambodia – formation process and regulations. *Journal of the Faculty of Agriculture,* Kyushu University, 55, 3, 21-26.

Hunt C. A. G., 2009. Carbon sinks and climate change: forests in the fight against global warming: e-book. Cheltenham: Edward Elgar.

Imhoff M. L., 1995. Radar backscatter and biomass saturation-Ramifications for global biomass inventory. *IEEE Transactions on Geoscience and Remote Sensing,* 33, 2, 511–518.

Imhoff M. L., Sisk T. D., Milne A., Morgan G., Orr T., 1997. Remotely sensed indicators of habitat heterogeneity: use of synthetic aperture radar in mapping vegetation structure and bird habitat. *Remote Sensing of Environment,* 60, 217–227.

Ince T., 2010. Unsupervised classification of polarimetric SAR image with dynamic clustering: An image processing approach. *Advances in Engineering Software,* 41, 636-646.

IPCC., 2007. The Physical science basis: summary for policymakers, Intergovernmental Panel on Climate Change 2007.

Islam Md. N., Kurttila M., Mehtatalo L., Pukkala T., 2010. Inoptimality losses in forest management decisions caused by errors in an inventory based on airborne laser scanning and aerial photographs. *Canadian Journal of Forest Research*, 40, 2427-2438.

Isoguchi O., Shimada M., Uryu Y., 2009. A preliminary study on deforestation monitoring in Saumatra island by PALSAR. *Proceedings of IEEE,* IGARSS 09.

JAFTA., 2010. Study project on information analysis and management for forest cover change assessment in the Mekong River basin, Cambodia. 9-21 (unpublished report).

Jensen J. R., 2000. Remote sensing of the environmental an earth resource perspective. Prentice Hall U. K.

Kaipainen, T., Liski, J., Pussinen, A., and Karjalainen, T., 2004. Managing carbon sinks by changing rotation length in European forests. *Environmental Science & Policy* 7(3), 205-219.

Kasischke E. S., Melack J. M., Dobson M. C., 1997. The use of imaging radars for ecological applications- A review. *Remote Sensing of Environment,* 59, 2, 141–156.

Kasischke E., Bourgeau-Chavez S., Laura L., Rober Allison R., Wyatt Kevin H., Waddington James M., Turetsky Merritt R., 2009. Effects of soil moisture and water depth on ERS SAR backscatter measurements from an Alaskan wetland complex. *Remote sensing of Environment,* 113, 1868-1873.

Kellomäki, S. and Leinonen, S., 2005. Management of European forests under changing climate conditions. Joensuu: University of Joensuu, Faculty of Forestry. Tiedonantoja 163.

Kenward T., Lettenmaier D. P., Wood E. F., Fielding E., 2000. Effects of Digital Elevation Model accuracy of hydrologic predictions. *Remote Sensing of Environment.* 74, 432-444.

Kenzo T., Furutani R., Hattori D., Kendawang J. J., Tanaka S., Sakurai K., Ninomiya I., 2009. Allometric equations for accurate estimation of above ground biomass in logged over tropical rainforests in Sarawak, Malaysia. *Journal of Forestry Research,* 14, 365-372.

Khatun K., 2011. Reconciling timber provision with carbon sequestration opportunities in the tropical forests of Central America. *Environmental Science and Policy,* doi: 10.1016/j.envsci.2011.05.018.

Kim Phat., Ouk N., Uozumi Y., Ueki T., 2001. A case study of current situation for forest concession in Cambodia-constraints and prospects. *Journal of Forest Planning,* 7, 2, 59-67.

Kim Y., Van Zyl J., 2001. Comparison of forest estimation techniques using SAR data, Proc. IGARSS2001.

Kiyono Y., Furuya N., Sum T., Umemiya C., Itoh E., Araki M., Matsumoto M., 2010. Carbon stock estimation by forest measurement contributing to sustainable forest management in Cambodia. *Japan Agricultural Research Quarterly,* 44, 1, 81-92.

Knuth R., Thiel C., Thiel C., Eckardt R., Richter N., Schmullius C., 2009. Multisensor analysis for forest monitoring in boreal and tropical forest environments. *Proceedings of IEEE IGARSS09.*

Kohavi R., Provost F., 1998. Glossary of Terms In: Editorial for the issue on applications of machine learning and the knowledge discovery process. *Machine learning,* 30, 2-3.

Kongrattanachok P., 2005. Carbon sequestration in cassava and para rubber plantation, Rayong province. Master thesis, Mahidol University.

Kummer D. M., Turner B. L., 1994. The human causes of deforestation in Southeast Asia. *Bioscience,* 44 (5), 323-326.

Kuntz S., 2010. Potential of spaceborne SAR for monitoring the tropical environments. *Tropical ecology,* 51, 1, 3-10.

Kuplich T. M., Curran P. J., Atkinson P. M., 2005. Relating SAR image texture to the biomass of regenerating tropical forests. *International Journal of Remote Sensing,* 26, 21, 4829-4854.

Lackey R. T., 1998. Seven pillars of ecosystem management. *Landscape and urban planning,* 40, 21-30.

Lal M., and Singh R., 2000. Carbon sequestration potential of Indian forests. *Environmental Monitoring and Assessment*, 60, 315-327.

Lal R., 2008. Carbon sequestration. Philosophical Transactions of the Royal Society B, 363, 815-830, 2008.

Lamb D., 2011. Regreening the bare hills: Tropical forest restoration in the Asia-Pacific region, Springer, New York.

Lamb D., Erskine P. D., Parrotta J. A., 2005. Restoration of degraded tropical forest landscapes. *Science,* 310, 5754, 1628-1632.

Larrazabal A. P., and Skutch M., 2011. A review of experience of community monitoring for REDD+. Input paper no. 2 for the FCPF workshop, Mexico City, 1-16.

Lavalle M., Solimini D., Pottier E., 2008. PolinSAR for forest biomass retrieval: PALSAR observations and model analysis. *Proceedings of IEEE,* IGARSS 08.

Le Toan T., Beaudoin A., Riom J., Guyon D., 1992. Relating Forest Biomass to SAR Data. *IEEE Transaction on Geoscience and Remote Sensing*, 30, 403–411.

Le Toan T., Quegan S., Davidson M.W.J., Balzter H., Paillou P., Papathanassiou K., Plummer S., Rocca F., Saatchi S., Shugart H., Ulander L., 2011. The BIOMASS mission: Mapping global forest biomass to better understand the terrestrial carbon cycle, *Remote Sensing of Environment,* doi:10.1016/j.rse.2011.03.020.

Le Toan T., Quegan S., Davidson M. W. J., Balzter H., Paillou P., Papathanassiou K., Plummer S., Rocca F., Saatchi S., Shugart H., Ulander L., 2011. The BIOMASS mission: Mapping global forest biomass to better understand the terrestrial carbon cycle. *Remote Sensing of Environment,* doi:10.1016/j.rse.2011.03.020.

Lee J. S., Ainsworth T. L., Chen K., 2008. Speckle filtering of dual polarization and polarimetric SAR data based on improved sifma filter. IEEE IGARSS, IV 21-24.

Lee J. S., Grunes M. R., Ainsworth T. L., Du Li., Schuler D. L., Cloude S. R., 1999. Unsupervised classification using polarimetric decomposition and the complex Wishart classification. *IEEE Transactions on Geoscience and Remote Sensing,* 37, 5, 2249-2258.

Lee J. S., Grunes M. R., Pottier E., Ferro-Famil L., 2004. Unsupervised terrain classification preserving polarimetric scattering characteristics. *IEEE Transactions on Geoscience and Remote Sensing,* 42, 4, 722-731.

Lee J. S., Pottier E., 2009. Polarimetric radar imaging: from basics to application, CRC Press, Taylor and Francis Group, Boca Raton.

Lee J. S., Wen J., Ainsworth T. L., Chen K., Chen A. J., 2009. Improved sigma filter for speckle filtering of SAR imagery. *IEEE Transactions on Geoscience and Remote Sensing,* 47, 1, 202-213.

Lefsky M. A., Cohen W. B., Harding D. J., Parker G. G., Acker S. A., Gower S. T., 2002. Lidar remote sensing of above-ground biomass in three biomes. *Global Ecology and Biogeography*, 11, 5, 393–399.

Lefsky M. A., Harding D. J., Keller M., Cohen W. B., Carabajal C. C., Del Bom EspiritoSanto F., Hunter M. O., de Oliveira Jr R., 2005. Estimates of forest canopy height and aboveground biomass using ICESat. *Geophysical Research Letters*, 32, L22S02.

Lin Y., Sarabandi K., 1999. Retrieval of forest parameters using a fractal-based coherent scattering model and a genetic algorithm. *IEEE Transactions on Geoscience and Remote Sensing,* 37, 3, 1415-1425.

Ling F., Chen Z., Wang Q., 2009. Comperison of ALOS PALSAR RVI and Landsat TM NDVI for forest area mapping. IEEE, APSAR 2009.

Liski, J., Palosuo, T., Peltoniemi, M. &Sievanen, R., 2005. Carbon and decomposition model Yasso for forest soils. *Ecological Modelling,* 189, 1-2, 168-182.

Lu D., 2006, The potential and challenge of remote sensing-based biomass estimation, *International Journal of Remote Sensing,* 27, 7, 1297-1328.

Lu D., 2005. Aboveground biomass estimation using Landsat TM data in Brazilian Amazon. *International Journal of Remote Sensing*, 26, 12, 2509-2525.

Lucas R. M., Mitchell A. L., Rosenqvist A., Proisy C., Melius A., Ticehurst C., 2007. The potential of L-band SAR for quantifying mangrove characteristics and change: Case studies from the tropics. *Aquatic Conservation - Marine and Fresh water Ecosystems,* 17, 245- 264.

Lucas R., Armston J., Fairfax R., Fensham R., Accad A., Carreiras J., Kelley J., Bunting P., Clewley D., Bray S., Metcalfe D., Dwyer J., Bowen M., Eyre T., Laidlaw M., Shimada M., 2010. An Evaluation of the ALOS PALSAR L-Band Backscatter—Above Ground Biomass Relationship Queensland, Australia: Impacts of Surface Moisture Condition and Vegetation Structure. Selected Topics in Applied Earth Observations and Remote Sensing, *IEEE Journal of 3, 4,* 576-593.

Lucas R.M., Cronin N., Lee A., Moghaddam M., Witte C., Tickle, P., 2006. Empirical relationships between AIRSAR backscatter and LiDAR-derived forest biomass, Queensland, Australia. *Remote Sensing of Environment,* 100, 407–425.

Luckman A., Baker J., Kuplich T. M., Yanasse C. D. F., Frery A. C., 1997. A study of the relationship between radar backscatter and regenerating tropical forest biomass for spaceborne SAR instruments. *Remote Sensing of Environment,* 60, 1, 1–13.

Luckman A., Baker J. R., Honzak M., Lucas R., 1998. Tropical Forest Biomass Density Estimation using JERS-1 SAR: Seasonal Variation, Confidence Limits and Application to Image Mosaics. *Remote Sensing of Environment,* 63, 126-139.

Macelloni G., Paloscia S., Pampaloni P., Marliani F., Gai M., 2001. The relationship between the backscattering coefficient and the biomass of narrow and broad leaf crops. *IEEE Transactions on Geoscience and Remote Sensing,* 39, 873–884.

MAFF., 2004. Agricultural Statistics 2003-2004. Phnom Penh, Cambodia, Ministry of Agriculture, Forestry and Fisheries.

MAFE., 2010. Ministry of agriculture, forestry and fisheries report. (http://www.stats.maff.gov.kh/en/index.php?page=stat&mode=rubberdried&option=com_content&Itemid=101).

Maire G., Marsden C., Nouvellon Y., Grinand C., Hakamada R., Stape J., Laclau J., 2011. MODIS NDVI time-series allow the monitoring of Eucalyptus plantation biomass. *Remote Sensing of Environment*, doi: 10.1016/j.rse.2011.05.017.

Mallinis G., Koutsias N., Makras A., Karteris M., 2004. Forest Parameters Estimation in a European Mediterranean Landscape Using Remotely Sensed Data. *Forest Science,* 50, 4, 450-460.

Maniatis D., 2010a. Metholologies to measure aboveground biomass in the Congo basin forest in a UNFCCC REDD+ context. Ph D thesis, University of Oxford.

Maniatis D., Mollicone D., 2010b. Options for sampling and stratification for national forest inventories to implement REDD+ under the UNFCCC. *Carbon Balance and Management,* 5, 9, 1-14.

Martin C. F., Thomas R. H., Krabill W. B., Manizade S. S., 2005. ICESat range and mounting bias estimation over precisely-surveyed terrain. *Geophysical Research Letters,* 32, L21S07.

McDonald K. C., Ulaby F. T., 1993. Radiative transfer modeling of discontinuous tree canopies at microwave frequencies. *International Journal of Remote Sensing,* 14, 11, 2097-2128.

McKenney, B. and Tola, P. 2002, Natural Resources and Rural Livelihoods in Cambodia: a Baseline Assessment. CDRI Working Paper No. 23. Phnom Penh.

Meridian Institute, 2009a. Reducing emissions from deforestation and forest degradation: an options assessment report. Prepared for the Government of Norway, by Angelsen A., Brown S., Loisel C., Peskett L., Streck, C., Zarin, D., http://www.REDD-OAR.org.

Meridian Institute, 2009b. REDD+ institutional options assessment. Prepared for the Government of Norway, by Streck C., Gomez-Echeverri L., Gutman P., Loisel C., Werksman J., http://www.REDD-OAR.org.

Mette T., Papathanassiou K., Hajnsek I., Zimmermann R., 2003. Forest biomass estimation using polarimetric SAR interferometry. Proceedings of IGARSS'03, July 21-25, 2003, Toulouse, Italy.

Mette T., Papathanassiou K., Hajnsek I., Pretzsch H., Biber P., 2004. Applying a common allometric equation to convert forest height from Pol-InSAR data to forest biomass. Proceedings of IGARSS'04, Sep. 20-24, 2004, Anchorage, USA.

Mitchard E. T. A., Saatchi S. S., Woodhouse I. H., Nangendo G., Ribeiro N. S., Williams M., Ryan C. M., Lewis S. L., Feldpausch T. R., Meir P., 2009. Using satellite radar backscatter to predict above-ground woody biomass: A consistent relationship across four different African landscapes. *Geophysical Research Letters,* 36, L23401, doi:10.1029/ 2009GL040692.

Mitchard E. T. A., Saatchi S. S., Lewis S. L., Feldpausch T. R., Woodhouse I. H., Sonke B., Rowland C., Meir P., 2011. Measuring biomass changes due to woody encroachment and deforestation/degradation in a forest–savanna boundary region of central Africa using multi-temporal L-band radar backscatter. *Remote Sensing of Environment,* doi:10.1016/ j.rse.2010.02.022.

Mougin E., Proisy C., Marty G., Fromard F., Puig H., Betoulle J. L., Rudant J. P., 1999. Multifrequency and multipolarization radar backscattering from mangrove forestes. *IEEE Transactions on Geoscience and Remote Sensing,* 37, 1, 94-102.

Muukkonen P., and Heiskanen J., 2007, Biomass estimation over a large area based on standwise forest inventory data and ASTER and MODIS satellite data: A possibility to verify carbon inventories. *Remote Sensing of Environment,* 107, 617-624.

Nabuurs G. J., Masera O, Andrasko K, Benitez-Ponce P, Boer R,Dutschke M, Elsiddig E, Ford-Robertson J, Frumhoff P, Karjalainen T, Krankina O, Kurz WA, Matsumoto M, Oyhantcabal W, Ravindranath N. H., Sanz Sanchez M. J., Zhang X. 2007. Forestry. In: Metz B, Davidson OR, Bosch PR, Dave R, Meyer LA, editors. Climate change 2007: mitigation. Contribution of Working Group III to the Fourth Assessment Report of the Intergovernmental Panel on Climate Change. Cambridge (UK) and New York, NY (USA): Cambridge University Press.541-584.

Naesset E., 1997. Determination of mean tree height of forest stands using airborne laser scanner data. *ISPRS Journal of Photogrammetry and Remote Sensing,* 52, 49–56.

Nilsson M., 1996. Estimation of tree heights and stand volume using an airborne lidar system. *Remote Sensing of Environment,* 56, 1, 1–7.

Ni-Meister W., Lee S., Strahler A. H., Woodcock C. E., Schaaf C., Yao T., Ranson J., Sun G., Blair J. B., 2010. Assessing general relationships between aboveground biomass and vegetation structure for improved carbon estimate form lidar remote sensing. *Journal of Geophysical Research,* 115, 1-12.

NIS, 2008. General population census of Cambodia 2008. Provisional population totals, National Institute of Statistics, Ministry of Planning, Phnom Penh, Cambodia.

Nizalapur V., Jha C. K., Masugundu R., 2010. Estimation of above ground biomass in Indian forested area using multi-frequency DLR-ESAR data. *International Journal of Geomatics and Geosciences,* 1, 2, 167-179.

Oberndorf R. B. J. D., 2006. Legal analysis of forest and land laws in Cambodia. *Community Forestry International*, report, 1-38.

Olander L. P., Gibbs H. K., Steininger M., Swenson J. J., Murray B. C., 2008. Reference scenarios for deforestation and forest degradation in support of

REDD: a review of data and methods. *Environment Research Letters*, 3, 1-11.

Omaliss K., 2010. REDD Development in Cambodia. *Powerpoint Presentation.* Tokyo, Japan.

Ouarzeddine M., Aissa A. B., Souissi B., Belkhider M., Boulahbal S., 2005. Polarimetric classification using the Cloude/Pottier decomposition. PolinSAR, 2005.

Ouarzedine M., Souissi B., Belhadj-Aissa A., 2009. Forest characterization and mapping using fully polarimetric SAR data. PolInSAR 2009, Italy.

Page S. E., Hoscilo A., Langner A., Tansey K., Siegert F., Limin S., 2009. Chapter 9: Tropical peatland fires in Southeast Asia. In M. A. Cochrane (Ed.), Tropical fire ecology: Climate change, land use, and ecosystems dynamics (263 − 287). Berlin: Springer-Praxis.

Palmann C., Mavromatis M., Sequeira J., Brisco B., 2008. Earth observation using radar data: an overview of applications and challenges. *International Journal of Digital Earth,* 1, 171-195.

Paloscia S., Macelloni G., Pampaloni P., Sigismondi S., 1999. The potential of C- and L-Band SAR in estimating vegetation biomass: The ERS-1 and JERS-1 experiments. *IEEE Transactions on Geoscience and Remote Sensing,* 37, 2107–2110.

Papathanassiou K. P., Cloude S. R., 2003. The Effect of Temporal Decorrelation on the Inversion of Forest Parameters from Pol-InSAR Data. IEEE, IGARSS-03.

Patel P., Srivastava H. S., Navalgund R. R., 2009. Use of synthetic aperture radar polarimetry to characterize wetland targets of Keoladeo National Park, Bharatpur, India. *Current Science,* 97, 4, 25, 529-537.

Patenaude G., Milne R., Dawson T. P., 2005. Synthesis of remote sensing approaches for forest carbon estimation: reporting to the Kyoto Protocol. *Environmental Science and Policy,* 8, 2, 161-178.

Paul, K. I., Polglase, P. J., Nyakuengama, J.G., Khanna, P.K., 2002. Change in soil carbon following afforestation. *Forest Ecol. Manage.* 168, 241–257.

Peter K. V., 2007. Commercial crops technology. New India Publishing Agency (New Delhi, India).

Phat K., Ouk N., Uozumi Y., Ueki T., 2001. A case study of the current situation for forest concessions in Cambodia. *Journal of Forest Planning,* 7, 2, 59-67.

Phelps J., Webb E.L., Agrawal A., 2010. Does REDD + threaten to recentralize forest governance? *Science,* 328, 312–313.

Phillips O. L., Malhi Y., Higuchi N., Laurance W. F., Percy V., Nunez P. V., Vasquez R. M., Laurance S. G., Ferreira L. V., Stern M., Brown S., Grace J., 1998. Changes in the carbon balance of tropical forests: evidence from long-term plots. *Science,* 282, 439–442.

ffenberger, M., 2006. People in the forest: community forestry experiences from Southeast Asia. *International Journal of Environment and Sustainable Development*, 5, 1.

Poffenberger M., 2009. Cambodia's forests and climate change: Mitigating drivers of deforestation. *Natural Resources Forum*, 33, 285–296.

Pohjola, J. and Valsta, L., 2007. Carbon credits and management of Scots pine and Norway spruce stands in Finland. *Forest Policy and Economics* 9, 7, 789-798.

Pottier E., 1998. Unsupervised classification scheme and topography derivation of PolSAR data based on the H/α/A polarimetric decomposition theorem. International workshop on Radar polarimetry, France, 535-548.

Pottier E., Ferro-Famil., Lee J. S., 2008. POLSARPRO V4.0 – lecture notes: Advanced concepts. [Internet] Available at: http://earth.esa.int/ polsarpro/.

Pratihast A. K., and Herold M., 2011. Community based monitoring and potential links with national REDD+ MRV. Input paper no. 1 for the FCPF workshop, Mexico City, 1-13.

Putz F. E., Zuidema P. A., Pinard M. A., Boot R. G. A., Sayer J. A., Douglas S., Sist P., Vanclay J. K., 2008. Improved tropical forest management for carbon retention. *PLoS Biol,* 6, 7, e166, 1368-1369.

Qi Z., Yeh A. G., Li X., Lin Z., 2010. Land use and land cover classification using RADARSAT-2 polarimetric SAR image. IAPRS, XXXVIII, part 7A, 198-203.

Ra, K., Pichdara, L., Dararath, Y., Jiao, X. Smith-Hall, C., 2011. Towards understanding household-level forest reliance in Cambodia-study sites, methods, and preliminary findings. Forest and Landscape, University of Copenhagen, working paper 60/2011, 7-11.

Rajaona A. M., 2008. Comparative study of allometric parameters of Cashew tree (anacardium occidentale) In North East Brazil, Master thesis, University of Bonn.

Ramankutty N., Gibbs H.K., Achard F., DeFries R., Foley J.A., Houghton R.A., 2007. Challenges to estimating carbon emissions from tropical deforestation. *Global Change Biology*. 13, 51-66.

Ranson K. J., Sun G., 1994a. Mapping biomass of a northern forest using multifrequency SAR data. *IEEE Transactions on Geoscience and Remote Sensing,* 32,388–396.

Ranson K. J., Sun G., 1994b. Northern forest classification using temporal multi frequency and multi-polarimetric SAR images. *Remote Sensing of Environment,* 47, 2, 142-153.

Ranson K. J., Sun G., 1997. An evaluation of AIRSAR and SIRC/X-SAR images for mapping northern forests attributes in Maine, USA. *Remote Sens Environ,* 59, 203–222.

Rauste Y., Hammer T., Pulliainen J., Hallikainen M., 1994. Radar-based forest biomass estimation. *International journal of remote sensing,* 15, 14, 2791-2808.

Ravindranath N. H., and Ostwald M., 2008. Carbon inventory methods, Handbook for greenhouse gas inventory, carbon mitigation and roundwood production projects. *Springer,* ISBN-13: 978-1-4020-6546-0.

REDD., 2010. Source book. GOFC-GOLD.

Ribbes F., Toan T.L., Bruniquel J., Floury N., Syussi N., Liew S.C., Wasrin U.R., 1997. Forest mapping in tropical region using multitemporal and interferometric ERS-1/2 data. In: Proceedings of the 3rd ERS symposium on space at the service of our environment, Florence, Italy, Estec, Noodrwijk, The Netherlands, 351–356 ESA SP-414.

Rignot E., Way J., 1994. Monitoring freeze-thaw cycles along north-south Alaskan transects using ERS-1 SAR. *Remote Sensing of Environment*, 49, 131–137.

Rignot E. J., Zimmermann R., Vanzyl J. J., 1995. Spaceborne applications of P band imaging radars for measuring forest biomass. *IEEE Transactions on Geoscience and Remote Sensing,* 33, 5, 1162-1170.

Rosenqvist A., Milne A., Lucas R., Imhoff M., Dobson C., 2003. A review of remote sensing technology in support of the Kyoto Protocol. *Environmental Science and Policy*, 6, 441–455.

Rudel T. K., Defries R., Asner G. P., Laurance W. F., 2009. Changing drivers of deforestation and new opportunities for conservation. *Conservation Biology,* 23, 6, 1396-1406.

Running, S. W., Nemani, R. R., 1988. Relating the seasonal pattern of the AVHRR normalized difference vegetation index to simulated photosynthesis and transpiration of forests in different climates. *Remote Sens. Environ.* 17, 472–483.

Running, S. W., Nemani, R. R., Peterson, D. L., 1989. Mapping regional forest evapotranspiration and photosynthesis by coupling satellite data with ecosystem simulation. *Ecology,* 70, 1090–1101.

Ruste Y., Hame T., Pulliainen J., Heiska K., Hallikainen M., 1994. Radar-based forest biomass estimation. *International Journal of Remote Sensing,* 15, 2797-2808.

Saatchi S. S., Harris N. L., Brown S., Lefsky M., Mitchard E. T. A., Salas W., Zutta B. R., Buermann W., Lewis S. L., Hagen S., Petrova S., White L., Silman M., Morel A., 2011. Benchmark map of forest carbon stocks in tropical regions across three continents, PNAS 2011, 108, 9899-9904, doi:/10.1073/pnas.1019576108.

Sader S. A., Waide R. B., Lawrence W. T., Joyce A. T., 1989. Tropical forest biomass and successional age class relationships to a vegetation index derived from Landsat TM data. *Remote Sensing of Environment,* 28, 143–156.

Sales M. H., Souza Jr C. M., Kyriakidis P. C., Roberts D. A., Vidal E., 2007. Improving spatial distribution estimation of forest biomass with geostatistics: A case study for Rondonia, Brazil. *Ecological Modelling,* 205, 1-2, 221-230.

Sandberg G., Ulander L. M. H., Fransson J. E. S., Holmgren J., Le Toan T., 2011. L- and P-band backscatter intensity retrieval in heniboreal forest. *Remote Sensing of Environment,* doi:10.1016/j.rse.2011.03.018.

Santoro M., Eriksson L., Askne L. J., Schmullius C., 2006. Assessment of standwise stem volume retrieval in boreal forest from JERS-1 L-band SAR backscatter. *International Journal of Remote Sensing*, 27, 16, 3425–3454.

Santos J. R., Freitas C. C., Araujo L. S., Dutra L. V., Mura J. C., Gama F. F., Soler L. S., Sant'Anna S. J. S., 2003. Airborne P-band SAR applied to the aboveground biomass studies in the Brazilian tropical rainforest. *Remote Sensing of Environment,* 87, 4, 482–493.

Sarabandi K., Lin Y. C., 2000. Simulation of interferometric SAR response for characterizing the scattering phase center statistics of forest canopies. *IEEE Transactions on Geoscience and Remote Sensing*, 38, 1, 115-125.

Sasaki N., Yoshimoto A., 2010. Benefits of tropical forest management under the new climate change agreement-a case study in Cambodia. *Environment Science and Policy,* 13, 384-392.

Sato T., 2011. Methods for biomass monitoring. International technical seminar on REDD+ knowledge sharing and discussion from practices and experiences, Tokyo, Japan 16-17 February, 2011. (http://www.ffpri.affrc.go.jp/redd-rdc/en/seminars/reports/2011/02/16/01/23_Tamotsu_Sato.pdf) access on 26/02/2011.

Saunier S., 2007. Final calibration/validation report – PRISM GAEL Consultant, Report Reference #GAEL-P237-DOC-007.

Schlamadinger B. and Marland G., 1996. The role of forest and bioenergy strategies in the global carbon cycle. *Biomass and Bioenergy*, 10, 5/6, 275-300.

Schlerf M., Atzberger C., 2006. Inverstion of a forest reflectance model to estimate structural canopy variables form hyperspectral remote sensing data. *Remote Sensing and Environment,* 100, 3, 281-294.

Schoups G., Troch P. A., Veroest N., 1998. Soil moisture influences on the radar backscattering of sugar beet fields. *Remote Sensing of Environment,* 65, 184-194.

Schrope M., 2009. When money grows on trees. *Nature Reports Climate Change,* 3, 101–103, doi:10.1038/climate.2009.78.

Schulze E. D., 2006. Biological control of the terrestrial carbon sink. *Biogeosciences,* 3, 147-166.

Shimada M., Isoguchi O., Tadono T., Isono K., 2009. PALSAR polarimetric calibration and geometric calibration. *IEEE, Transactions on Geoscience and Remote Sensing*, 47, 12, 3915-3932.

Shimada M., 2010. Ortho-rectification and slope correction of SAR data using DEM and its accuracy evaluation. *IEEE Journal of Selected Topics in Applied Earth Observations and Remote Sensing,* 3, 4, 657-671.

Shono K., Cadaweng E. A., Durst P. B., 2007. Application of assisted natural regeneration to restore degraded tropical forestlands. *Restoration Ecology,* 15, 4, 620-626.

Simard M., Rivera-Monroy V.H., Ernesto Mancera-Pineda J., Castañeda-Moya E., Twilley R.R., 2008. A systematic method for 3D mapping of mangrove forests based on shuttle radar topography mission elevation data, ICEsat/GLAS waveforms and field data: Application to Ciénaga Grande de Santa Marta, Colombia. *Remote Sensing of Environment,* 112, 5, 2131–2144.

Singh G., Yamaguchi Y., Park S. E., 2011. Utilization of four component scattering poser decomposition method for glaciated terrain classification. *Geocarto International*, doi: 10.1080/10106049.2011. 584978.

Sivanpillai R., Smith C. T., Srinivasan R., Messina M. G., Wu X. B., 2006. Estimation of managed loblolly pine stand age and density with Landsat ETM+ data. *Forest Ecology and Management*, 223, 247-254.

Skutsch M., 2010. Community forest monitoring for the carbon market: opportunities under REDD. Earthscan, London.

Skutsch, M. M. and Ba, L., 2010. Crediting carbon in dry forests: The potential for community forest management in West Africa. *Forest Policy and Economics,* 12, 4, 264-270.

Skutsch, M. and Trines, E., 2011. The policy context of community monitoring for REDD+. In M. Skutsch, ed. Community forest monitoring for the carbon market. Opportunities under REDD. London, Washington DC: Earthscan, 43-60.

Steininger M. K., 2000. Satellite estimation of tropical secondary forest aboveground biomass: Data from Brazil and Bolivia. *International Journal of Remote Sensing*, 21, 6, 1139–1157.

Stephens S. S., Wanger M. R., 2007. Forest plantation and biodiversity: A fresh perspective. *Journal of Forestry*, 105, 6, 307-313.

Streck C., Gomez-Echeverri L., Gutman P., Loisel C., Werksman J., 2009. REDD+ Institutional Options Assessment: Developing an Efficient, Effective, and Equitable Institutional Framework for REDD+ under the UNFCCC. Meridian Institute. Washington DC, United State of America.

Sun G., Ranson K. J., Kimes D.S., Blair J.B., Kovacs K., 2008. Forest Vertical Structure from GLAS: An Evaluation Using Lvis and Srtm Data. *Remote Sensing of Environment,* 112, 1, 107-117.

Sun G., Ranson K. J., Guo Z., Zhang Z., Montesano P., Kimes D., 2011. Forest biomass mapping from lidat and radar synergies. *Remote Sensing of Environment,* doi:10.1016/j.rse.2011.03.021.

Suzuki K., Ishii K., Sakurai S., Sasaki S., 2006. Plantation technology in tropical forest science. *Springer,* 53-66.

Takaku K., Futamra N., Iijima T., Tadono T., Shimada M., Shibasaki R., 2004. High resolution DEM generation from ALOS PRISM data – simulation and elevation. IGARSS, 2004.

TEEB., 2010. The Economics of Ecosystems and Biodiversity for business.

Trisasongko B. H., 2010. The use of polarimetric SAR data for forest disturbance monitoring. *Sens Imaging,* 11, 1-13.

Tyler S., 2006. Communities, livelihoods, and natural resources: Action research and policy change in Asia. *Intermediate Technology Publications,* ISBN 1-85339-638-9.

Trotter C. M., Dymond J. R., Goulding C. J., 1997. Estimation of timber volume in a coniferous plantation forest using Landsat TM. *International Journal of Remote Sensing,* 18, 2209-2223.

Ulaby F. T., Moore R. K., Fung A. K., 1982. Microwave Remote Sensing: Active and Passive. Volume 2: Radar Remote Sensing and Surface

Scattering and Emission Theory, Addison- Wesley, Advanced Book Program: Reading, Massachusetts, 609.

UNEP., 2009. Cambodia environment outlook, Ministry of Environment, Kingdom of Cambodia.

Van Beukering, P. J. H., Leeuw, K. van der, Grogan, K., Hansfort, S. L., 2009. Reduced Emission from Deforestation and Degradation in the Southern Cardamom Ecosystem, Cambodia. IVM Report (R-09/13). Institute for Environmental Studies, VU University, Amsterdam.

Van Der Werf, G. R., Morton, D. C., De Fries, R. S., Olivier,J. G. J., Kasibhatla, P. S., Jackson, R. B., 2009. CO_2 emissions from forest loss. *Nature Geoscience,* 2, 737–738.

VanZyl J., 1993. The effect of topography on radar scattering from vegetated areas. *IEEE Transactions on Geoscience and Remote Sensing*, 31, 1, 153-160.

Wang G., Zhang M., Gertner G. Z., Oyana T., McRoberts R. E., Ge H., 2011. Uncertainties of mapping aboveground forest carbon due to plot locations using national forest inventory plot and remotely sensed data. *Scandinavian Journal of Forest Research,* 26, 360-373.

Wang H., Huang Q., Pi Y., 2009. Polarization decomposition with S and T matrix of a PolSAR image. ICCCAS, 2009.

Wang Q., Adiku S., Tenhunen J., Granier A., 2005. On the relationship of NDVI with leaf area index in a deciduous forest site. *Remote Sensing of Environment,* 94, 244–255.

Wang Y., Kasischke E. S., Melack J. M., Davis F. W., Christensen N. L., 1994. The effects of changes in Loblolly pine biomass and soil moisture on ERS-1 SAR backscatter. *Remote Sensing of Environment,* 49, 25–31.

Watanabe M., Shimada M., Rosenqvist A., Tadono T., Matsuoka M., Romshoo A.A., Ohta K., Furuta R., Nakamura K., Moriyama T., 2006. Forest structure dependency of the relation between L-band σ^o and biophysical parameters, *IEEE Transactions on geosciences and remote sensing,* 44, 11, 3154-3165.

Watson R. T., Noble I. R., Bolin B., Ravindranath N. H., Verardo D. J., Dokken D. J. (eds)., 2000. Land use, land-use change, and forestry. A special report of IPCC. Cambridge University Press, Cambridge.

Werf G. R., Morton D. C., Defries R. S., Olivier J. G. J., Kashibhatla P. S., Jackson R. B., Collatz G. J., Randerson J. T., 2009. CO_2 emissions form forest loss. *Nature Geocsience,* 2, 737-738.

Wijaya A., 2010. Complex land cover classifications and physical properties retrieval of tropical forests suing multi-source remote sensing. Ph D thesis, The Technische Universitat Bergakademie Freiberg.

Xiao X. M., Boles S., Frolking S., Li C. S., Babu J. Y., Salas W., Moore B., 2006. Mapping paddy rice agriculture in South and Southeast Asia using multi-temporal MODIS images. *Remote Sensing of Environment,* 100, 1, 95–113.

Yamaguchi Y., Moriyama T., Ishido M., Yamada H., 2005. Four-component scattering model for polarimetric SAR image decomposition, *IEEE Transaction on Geoscience Remote Sensing*, 43, 8, 1699-1706.

Yamaguchi Y., Yajima Y., Yamada H., 2006. A four-component decomposition of POLSAR images based on the coherency matrix. *IEEE Geoscience and Remote Sensing Letter,* 3, 292–296.

Yamanokuchi T., Doi K., 2007. Combined Use of InSAR and ICESat/GLAS Data for High Accuracy DEM Generation on Antarctica. *IEEE International Symposium on Geoscience and Remote Sensing,* doi: 10.1109/IGARSS.2007.4423028.

Zebker H. A., Van Zyl J. J., Durden S. L., Norikane L., 1991. Calibrated imaging radar polarimetry: Technique, examples, and applications. *IEEE Transaction on Geoscience and Remote Sensing,* 29, 6, 942–961.

Zhao K., Popescu S., Nelson R., 2009. Lidar remote sensing of forest biomass: A scale-invariant estimation approach us airborne lasers, *Remote sensing of environment,* 113, 182-196.

ABOUT THE AUTHOR

Ram Avtar
Research Fellow
United Nations University, Institute for the Advanced Study of Sustainability
(UNU-IAS), Japan

Dr. Ram Avtar currently working as Research Fellow at United Nations University, Institute for the Advanced Study of Sustainability (UNU-IAS), Japan. His research focuses on assessing the impact of global climate change and weighing on adaptation and mitigation options to offer scientific solutions. He has been working on various environmental issues related to climate change adaptation and mitigation. Dr. Avtar is recipient of the Green Talent Award-2013 by Federal Ministry of Education, Germany. He received his PhD from the University of Tokyo and explored gaps in existing biomass estimation practices and provided practical and policy relevant solutions to strengthen prevalent practices. Prior to joining United Nations University, Institute for the Advanced Study of Sustainability (UNU-IAS), Tokyo, Japan, Dr. Avtar worked as Post-Doctoral Researcher with the Japan Agency for Marine-Earth Science and Technology (JAMSTEC). He has completed his Master degrees (Master of Science and Master of Philosophy) from Jawaharlal Nehru University, India. Dr. Avtar has presented his research in numerous international conferences and extensively published papers in peer-reviewed international journals as well as books. His research interest includes application of remote sensing in forestry and REDD+ related issues and he sees greater potential in space technology to offer sustainable solutions to pressing environmental problems of the world. He is also involved in IPBES assessment report as a Lead Author.

INDEX

D

E

F

G

H

I

O

P

Q

R

S

T

U

V

W

Y